Measuring Construction

Despite the size, complexity and importance of the construction industry, there has been little study to date which focuses on the challenge of drawing reliable conclusions from the available data. The accuracy of industry reports has an impact on government policy, the direction and outcomes of research and the practices of construction firms, so confusion in this area can have far reaching consequences.

In response to this, *Measuring Construction* looks at fundamental economic theories and concepts with respect to the construction industry, and explains the merits and shortcomings, sometimes by looking at real life examples. Drawing on current research the contributors tackle:

- industry performance
- productivity measurement
- construction in national accounts
- comparing international construction costs and prices
- comparing international productivity.

The scope of the book is international, using data and publications from four continents, and tackling head on the difficulties arising from measuring construction. By addressing problems that arise everywhere from individual project documentation, right up to national industrial accounts, this much-needed book can have an impact at every level of the industry. It is essential reading for postgraduate construction students and researchers, students of industrial economics, construction economists and policy-makers.

Rick Best is Associate Professor of Construction Management and Economics at Bond University. He has produced numerous book chapters and papers over a 20 year career as an academia as well as co-editing three books and co-authoring two quantity surveying textbooks. He was the founding director of the Centre for Comparative Construction Research. His research over the past ten years has been focused on the problems associated with making valid comparisons of construction industries across countries and has contributed to the development of the most recent construction data collection project within the International Comparison Program.

Jim Meikle has a part-time chair in construction economics at the Bartlett School of Construction and Project Management, University College, London (UCL). He also works as an independent consultant, having retired as a partner of Davis Langdon LLP in 2005. He has extensive experience in the management and implementation of construction, consultancy and research projects and programmes and has worked for private clients and government departments in the UK and other countries, and for international organizations. His interests include international comparisons, the structure and operation of construction industries and construction industry statistics. Jim is a non-executive director with Alexi Marmot Associates (AMA), a trustee of the Usable Buildings Trust and an adjunct professor at Bond University.

Measuring Construction
Prices, Output and Productivity

Edited by
Rick Best and Jim Meikle

Routledge
Taylor & Francis Group

LONDON AND NEW YORK

First published 2015
by Routledge
2 Park Square, Milton Park, Abingdon, Oxon OX14 4RN

and by Routledge
605 Third Avenue, New York, NY 10017

First issued in paperback 2020

*Routledge is an imprint of the Taylor & Francis Group, an informa
business*

British Library Cataloguing-in-Publication Data
A catalogue record for this book is available from the British Library

Library of Congress Cataloging-in-Publication Data
Measuring construction : prices, output and productivity / edited by
  Jim Meikle, Rick Best. — 1 Edition.
    pages cm
  Includes bibliographical references and index.
 1. Construction industry.   I. Meikle, Jim.   II. Best, Rick.
  HD9715.A2.M3566 2015
  338.4'7624—dc23
  2014039295

ISBN 13: 978-0-367-73834-1 (pbk)
ISBN 13: 978-0-415-65937-6 (hbk)

Typeset in Sabon
by Apex CoVantage, LLC

To Cheryl and Sheilah

Contents

Acknowledgements

We would like to thank a number of people who have contributed in one way or another to the production of this book.

The contributing authors – without their effort and patience, there would not be a book.

The other people, apart from the contributors, who attended the initial workshop at Bond University in May 2012 – their contributions to the discussion helped to widen our views of many of the issues related to 'measuring construction'.

Richard Jedryas – who produced the figures and tables.

Alice Aldous and Matthew Turpie at Taylor & Francis – who piloted us through the process of preparing the book for publication.

Preface

The genesis of this book lies in a research workshop held at Bond University on Australia's Gold Coast in May 2012, organized by the Centre for Comparative Construction Research (CCCR). The CCCR had been established at Bond in the previous year by the editors of this book; it grew out of work that we had been engaged in for the World Bank in regard to construction cost comparison methodologies. The workshop brought together a number of individuals with an interest in the areas of construction cost and industry performance (see *Past Projects* at www.bond.edu.au/cccr for details).

The workshop included a number of leading construction economists from both academia and industry as well as representatives from the Australian Bureau of Statistics, a mainstream economist and a couple of engineers. The participants came from the US, the UK, Iran and New Zealand as well as Australia. It was a major event in our collaboration, which had begun quite informally in 2004 when Rick was working on a thesis based on international construction cost comparisons and met Jim, who was still at that time the head of Davis Langdon Consultancy in London. In 2009, on the basis of his more than 30 years of involvement with construction cost research for Eurostat and others, Jim was asked by the World Bank to review the methods being used for collecting and analysing cost data that is used in the construction component of the International Comparison Program. He invited Rick to be part of the team for this review, and this in turn led to the establishment of the CCCR.

We had previously discussed the idea of pooling our experience in a book and felt that bringing together some like-minded people would not only provide a forum in which we could explore and expand our ideas but also perhaps bring new players into the game. Two days of presentations and lively discussion did give the project impetus and did, indeed, add some new names to the list of authors who have contributed to this book. In the end, workshop participants contributed to all but one of the chapters.

The book covers a range of topics related to measurement in a fairly broad sense as it is applied to construction without dealing in any way with the technical skill of building measurement as it is generally understood and practised by quantity surveyors and estimators. While the range of topics is broad, it is far from complete, and there is still plenty to be explored that is not addressed here; perhaps a second volume will be required not only to deal with topics not covered but also to capture further developments in a field of research that is far from exhausted. We hope that the book will not only inform readers but also promote further discussion, as there is still work to be done to bring the methods used in comparative studies to maturity.

Note: In this book the word 'data' is treated as a collective noun and thus it takes the singular form as in '. . . the data is reliable . . .' rather than the somewhat archaic '. . . the data are reliable . . . .' While in Latin the singular is 'datum' and the plural 'data', this book is written in English, not Latin, and the editors are not aware of any common usage of the word 'datum' to describe or identify a single piece of data.

Rick Best – Gold Coast
Jim Meikle – London
August 2014

Contributors

Malcolm Abbott is an Associate Professor of Economics at the Swinburne University of Technology in Melbourne and has a PhD from the University of Melbourne. One of his main areas of expertise is productivity analysis, and he has published extensively on the productivity of utilities (electricity and gas), hospitals, airports and the construction industry in Australia and New Zealand. His other fields of research include the development of energy markets, utility regulation and the economics of education.

Rick Best is Associate Professor of Construction Management and Economics at Bond University. He has produced numerous book chapters and papers over a 20-year career as an academic as well as coediting three books and coauthoring one quantity surveying textbook. He was the founding director of the Centre for Comparative Construction Research. His research over the past 10 years has been focused on the problems associated with making valid comparisons of construction industries across countries and has contributed to the development of the most recent construction data collection project within the International Comparison Program.

Gerard de Valence is a Senior Lecturer in the School of the Built Environment in the Faculty of Design, Architecture and Building at the University of Technology, Sydney. Prior to becoming an academic in 1992, he had 10 years' experience as an analyst and economist in the private sector doing research on the property, building and construction industries for the Australian Stock Exchange, the Property Council of Australia and the NSW Royal Commission into Productivity in the Building Industry. He has a long-standing interest in industry performance and development. His research has broadly focused on issues around the structure, conduct and technological trajectory of the building and construction industry, with more than a hundred refereed papers and book chapters published.

Stephen Gruneberg is Reader in Construction Economics in the Faculty of Architecture and the Built Environment at the University of Westminster.

He has written several books on the economic theory of the construction sector. His most recent work was on the procurement of the London 2012 Olympics, which he wrote with John Mead. His research interests include global construction data and international aspects of construction. He is the coordinator of the CIB Task Group on Global Construction Data, a member of the Consultative Committee on Construction Industry Statistics (CCCIS) in the Department for Business Innovation and Skills (BIS) and, until recently, he was responsible for international liaison for the Association of Researchers in Construction Management.

Graham Ive is Senior Lecturer in the Economics of Construction at the Bartlett School of Construction and Project Management, University College London. His research has focused on aspects of the industrial economics of the construction sector, embracing the structure of the construction industries, the strategies, behaviour and performance of construction firms, and the complex and specific economic institutions of the construction process, specifically contracting and procurement systems.

Craig Langston is Professor of Construction and Facilities Management at Bond University in Queensland. In addition to authoring more than 100 papers, five books and three software programs, Professor Langston has been awarded four Australian Research Council (ARC) grants totalling about $1 million, won the Chartered Institute of Building International Innovation and Research Award (2013), Outstanding Paper for the *Facilities* journal (2013), Emerald Literati Network Outstanding Paper Award (2013) and Bond University Vice Chancellor's Quality Award for Research Excellence (2010). He has 30 years' experience as an academic, teaching a range of subjects related to the built environment. He is also Director of the Centre for Comparative Construction Research at Bond.

Weisheng Lu is Assistant Professor in the Department of Real Estate and Construction, Faculty of Architecture at the University of Hong Kong. He publishes widely on international construction topics such as competitiveness, SWOT analysis of construction business, cross-jurisdiction comparative studies and corporate social responsibility (CSR) in international construction. By working closely with Professor Roger Flanagan at the University of Reading, UK, he is shifting his research focus from traditional contracting business to construction professional services.

Jim Meikle retired as a partner of Davis Langdon LLP in 2005. Since then, he has operated as an independent consultant working with UK government departments and agencies, foreign governments, international organizations including the EU, the World Bank and the African Development Bank, and private clients. His main interests are construction industry and research policy, construction data, construction productivity, construction professional services and international comparisons. Jim has a part-time chair in construction economics at the Bartlett School of

Construction and Project Management, University College London, and is an adjunct professor at Bond University, Queensland, Australia. He is also a nonexecutive director of AMA Alexi Marmot Associates and a trustee of the Usable Buildings Trust.

Ali Najafi received degrees of BSc in civil engineering in 2006 from Sharif University of Technology, Iran, MBA with specialization in IT management in 2009 from Multimedia University, Malaysia and PhD with specialization in construction management in 2014 from Nanyang Technological University, Singapore. He is a registered professional engineer in Iran. His main research interests and publications are in the field of productivity modelling, IT applications in construction management and building information modelling (BIM). As part of his PhD studies, he worked with several contractors as well as the Singapore Housing and Development Board to develop a web-based system for productivity modelling of precast concrete installation times. He is currently working as a productivity specialist for an international contracting firm in Iran.

Göran Runeson is currently Adjunct Professor at the University of Technology, Sydney. Recently retired, he divides his time between editing the *Australasian Journal of Construction Economics and Building* and supervising PhD students. Dr Runeson started his academic career as a lecturer in economics before transferring to construction management. He is the author of three books and some 200 publications on various aspects of building economics and research methodology.

Robert Tiong is Associate Professor, School of Civil and Environmental Engineering, Deputy Director, Centre for Infrastructure Systems (2006–2011), the Institute of Catastrophe Risk Management (2011–2013) and BIM Center of Excellence (2013–Current), Nanyang Technological University, Singapore. He holds a bachelor of civil engineering with specialization in management from University of Glasgow, master of engineering in construction management from University of California, Berkeley, USA and PhD from Nanyang Technological University, Singapore. He is a registered professional engineer in Singapore. His research interests and publications are in build-operate-transfer (BOT) projects, public–private partnership for infrastructure projects, international project finance, productivity in construction and building information modelling. He has worked with the Asian Development Bank and the World Bank on risk analysis and management of infrastructure projects.

Huan Yang is Lecturer in the Urban and Regional Science Faculty at Shanghai University of Finance and Economics, PRC. She graduated from a joint PhD program between Southeast University (China) and the Hong Kong Polytechnic University in 2010, with her major in management science and engineering. Her research interests include international business, strategic management, and organizational ecology, especially for companies from the construction and manufacturing industries. Her research findings have been published in internationally recognized journals such as *ASCE Journal of Construction Engineering and Management, ASCE Journal of Management in Engineering* and *Construction Management and Economics*.

Marco Yu recently completed his PhD at the Bartlett School of Construction and Project Management, University College London. He read economics at Cambridge and is a chartered surveyor. His research interests are in construction economics and statistics.

1 Setting the scene

Jim Meikle and Rick Best

Introduction

Measurement is the first step that leads to control and eventually to improvement. If you can't measure something, you can't understand it. If you can't understand it, you can't control it. If you can't control it, you can't improve it.
H. James Harrington

Peter Drucker, a prolific writer on various aspects of management, is often erroneously cited as the originator of the phrase, "If you can't measure it, you can't manage it". Exactly where the phrase came from is unknown, but the idea remains valid. Measuring construction in the sense of measuring quantities of building work is a well-established technical process carried out routinely by quantity surveyors and estimators, and in many places there are agreed-upon rules and procedures that are used for such work. Measuring construction industries or parts of those industries is another matter. While there must have been construction managers as far back as the Pyramid Project, the discrete discipline of construction management is relatively new, and serious scientific research in the area is still evolving. Measuring construction, in the sense of measuring things such as performance and productivity, is still the subject of some of that research, and methods for doing it are far from being fully developed and generally accepted by all interested parties.

The measurement of physical construction work gave rise to the emergence of quantity surveyors as a separate profession; the term 'surveyor' was used because people were required to go to sites and physically 'survey' or measure the quantum of work completed: lengths of walls of a certain height and thickness, for example, in order that the builder could be paid. Today there are standard methods for the measurement of physical building work, but the same cannot be said for the measurement of the characteristics of the construction industry. Even measuring construction cost can be done in a variety of ways, particularly if the aim is to measure cost, say, in a comparative exercise where costs in one place or country are compared to costs in another. This becomes more complex when the various costs are recorded in different currencies and/or different points in time.

Comparative studies

Industry comparisons are a primary focus of this book; for more than 60 years, researchers in academia and government have been trying to assess the relative performance of construction industries across countries. The initial aim is usually to measure various aspects of industry performance in order to compare the results in one place with those from another and thus to identify shortcomings in one place and then to look for ways to deal with them. If, for example, research in country A suggests that it is cheaper to build in country B, the natural question that follows: why is it cheaper? Is it because inputs such as materials and labour are cheaper, or is it that the way projects are set up and managed in country B leads to greater efficiency in the construction process and thus to reduced costs overall? Even finding a universally acceptable method for arriving at the conclusion that construction in country B is 'cheaper' is a major challenge, as assessing relative construction costs is far from an exact science, and there is no generally agreed method for doing it.

At face value, it appears to be a simple enough question that could be easily addressed by, for example, working out what it would cost to build a typical building in each country and then comparing the costs. In reality, it is quite a complex problem which has no simple answer. At the macro level, there are two major complications: one is that even if it were possible to devise a perfect method which was unanimously endorsed as the right way to make such comparisons, it is unlikely, for a number of reasons, that such a method could be successfully implemented. Second, that there is no 'correct' answer against which results can be compared, so it is impossible to know whether the method or the outcomes are robust and reliable.

At a detailed level, the problems multiply; some of the difficulties are:

- What is a 'typical' building? Even if we select a common type of building, say, a simple open-plan warehouse, local conditions, regulations, conventions and client expectations (among other things) will probably mean that the buildings may be functionally similar and even physically similar but far from identical.
- Is such a 'typical' building representative of the whole construction industry in a location, be it a city, region or whole country?
- It is most unlikely that buildings that are sufficiently similar in type and size will be constructed in different locations on similar sites at the same time.

Since 1950, numerous attempts have been made to compare aspects of construction industry performance between countries. These include studies using varying methodologies based on single projects, real (e.g. Lynton 1993) or hypothetical (e.g. Xiao and Proverbs 2002) and multiple projects, real (e.g. Flanagan *et al.* 1986; Langston and Best 2000) or hypothetical (e.g. Langston and de Valence 1999; DLC 1999). Edkins and Winch (1999) reviewed many of the studies carried out in the latter part of the 20th century and outlined the variety of approaches adopted by various researchers in the search for robust methods for assessing how the UK construction industry was performing.

Some studies were based on functionally similar buildings (e.g. Freeman, 1981) where buildings that served the same purpose, rather than being physically identical, were used; the idea was to ascertain what it would cost to construct a building according to local requirements to satisfy a particular need, such as a single-family house that provides shelter and security for a family with average income. Such a house, however, in southern Italy, for example, is not likely to bear much resemblance to a house fulfilling the same need in Norway and, apart from the design of their houses, would an 'average-income family' in the two locations actually be similar in terms of size, income and so on. The potential differences are obvious, but do those differences make the comparison any less valid than pricing an identical house in both locations? Pricing a Nordic house in southern Italy would involve pricing extensive thermal insulation and central heating that would be unlikely inclusions in a Mediterranean house, so such a comparison may have little meaning.

Industry performance

Various stakeholders are interested in assessing the performance of the construction industry in their city, region or country and in comparing that performance to that of comparable industry sectors in other places. In the UK, for example, the Latham and Egan reports have been drivers for change in their construction industry; in Australia, governments (state and national) as well as interested groups such as the Business Council of Australia (BCA) have a keen interest in the efficiency and productivity of the construction industry. In 2012, the BCA published a major report in which it was claimed that Australia was a high-cost, low-productivity place to build infrastructure compared to the US (BCA 2012). The results of this study were reported in the media with a hint of sensationalism (Hepworth 2012; Forrestal and Dodson 2012), and it caught the attention not only of the government but also of various industry bodies such as the Australian Constructors Association. Best (2012) demonstrated that two aspects of the BCA study were flawed – one related to the data used and the other to the method that was used to compare AUD and USD costs. The Australian government subsequently directed the Productivity Commission to explore the broader questions of how public infrastructure is procured and how long lead times and high costs associated with such projects could be reduced (Productivity Commission 2014).

Data issues

Measurement of construction at an industry or sectoral level involves two main elements: data and methods. In this book, both are discussed in the contexts of construction output, prices, performance and productivity. Collecting reliable construction data is always problematic, and it is particularly difficult to observe construction activity, construction employment and construction costs or prices. Construction is carried out in projects; these are diverse in size, type and complexity and are always site specific to some degree.

Data is typically based on partial information and/or estimates. For example, measurement of construction activity should include new build as well as renovation and maintenance as well as do-it-yourself (DIY), informal and cash economy work that is not included in statements of income by those who carry out the work. While new construction of housing and major projects may be well recorded, much of the work in other categories is unrecorded and must be either estimated (a difficult task given the lack of information available) or ignored (a poor option, as a significant proportion of total activity may then be missed). In some developing countries, informal construction may account for anything up to 90% of construction activity (Meikle, 2011), yet quantifying such work with any accuracy is fraught with problems, not the least of which is how to express the quantum of work (usually measured in terms of monetary value) when much of the work is done by householders often assisted by local people on the basis of community spirit and goodwill with no money changing hands for labour inputs and materials sourced directly from the locale (e.g. mud bricks), by barter or second hand.

Other measures of activity that are typical of developed countries, involving concepts such as gross output, value added and contractors' output, are interpreted differently, and thus data collected in different countries is often not directly comparable, as the data simply may not represent the same thing in different countries. Construction employment is similarly problematic, as it may or may not include casual workers and self-employed people or those carrying out DIY work. In fact, even placing boundaries on what a 'construction industry' includes is far from clear; for example, professional services such as project management and design fees may or may not be included. It is for this reason that any statement along the lines of 'construction represents 10% of GDP' must be treated with caution unless the parts of the broader industry that are included or excluded are clearly defined. If it is 'contractors' output', it will be a different figure than one that includes fees paid by clients, although in many design and build (D&B) contracts, design costs are included in construction contracts. Rawlinsons (2014) suggest, for example, that in Australia, professional fees may represent somewhere between 6.5% and 17% of project cost depending on the size and type of project.

Price (or cost) data may be out-turn costs (the total price finally paid by the client to 'purchase' construction work), input costs (prices paid by contractors for resources that include materials, components, labour and equipment) or tender or contract sum (the amount that the client agrees to pay to the contractor when the contract is let – this will, however, seldom be the same as the eventual out-turn cost). All these values have their attendant difficulties. True out-turn prices are seldom available but are generally different from project estimates and may vary considerably from the original contract sum once all variations, extensions of time and the like are added.

Input prices have a different set of constraints; list prices for materials and equipment hire often conceal major variations as a result of market conditions and the buying power of purchasers. Major contractors may obtain (or major projects may be eligible for) significant bulk discounts and/or loyalty discounts that suppliers offer in the hope of securing repeat business from large customers. Conversely, smaller firms or projects will often pay list price or above for what they buy. Similarly, in times of high demand, suppliers will raise prices and/or demand payment in advance, while in times of low demand, discounts and favourable credit terms will be more easily obtained.

Labour costs may be quoted from basic wage award agreements all the way through to all-in charge-out rates that include all allowances, oncosts and contractor's margin; the difference may be as much 3 to 400%. For example, according to Cordell (2014), the minimum award rate for a licensed plumber in Queensland, Australia, is AUD21.39/hour, while the same source quotes AUD68.00/hour for a subcontracted plumber. Rawlinsons (2014) suggest a charge-out rate including profit and overheads of AUD73.00 to 84.00/hour. For small domestic jobs, this rate (in April 2014) can approach AUD100.00/hour or more.

Material and labour prices can also vary considerably within countries, particularly physically large countries. The Bureau of Labor Statistics (2014) gives 2013 US hourly wage estimates for construction labourers ranging from USD9.46/hour to USD28.32/hour, with the hourly mean wage for labourers in New York (USD23.23) nearly double that in Texas (USD13.01). Ready-mixed concrete shows smaller but still significant regional price variations: according to RSMeans (2014), one cubic metre of 3000psi (approx. 20MPa) concrete in Washington, DC, costs USD154, which is around 25% more than the cost of the same item in Houston (USD123). The national average price is USD130/m^3.

In all of these cases, margins of error or uncertainty can be high, and when datasets are combined, for example, when productivity is measured using data on both output and employment, the problems are compounded. In this book, various authors recognize these issues and, where possible, offer at least partial solutions such as gathering multiple observations of prices, or triangulation or statistical methods.

Method issues

Even when good data is available, the methods used in working with it are not straightforward. Comparisons of price, for example, over time or across countries, require costs to be adjusted using price indices or currency normalizers. Using commercial or money market exchange rates (which economists refer to as 'nominal exchange rates') as the single conversion factor for all components in an economy is as inappropriate as using general inflation figures to normalize housing prices or overall construction costs over time.

The idea of using price indices to bring costs from different times to a common base date is familiar to most, but the idea of using currency normalizers rather than commercial exchange rates is not so well understood. As a result, there is one chapter in this book that deals with price indices, but there are a number that address international construction cost comparisons in general and the problem of currency conversion in particular. More specifically, the concept, formulation, production and use of purchasing power parities (PPPs) as an alternative to exchange rates is discussed in detail. The development of PPPs is not in itself an exercise in assessing the performance of construction or any other industry sector; PPPs are simply a neutral way of stating costs recorded in various national currencies in a single currency so that valid comparisons can be made. While costs are typically converted to US dollars, there are other noncurrency options that include 'construction dollars' and even hamburgers; these are discussed in later chapters.

It must be said, however, that no matter how good the method used, poor data will produce poor results. Discussions of method, if they are to produce anything that is useful in a practical sense, must include considerations of implementation, particularly in regard to the existence and availability of sound data. Even where data does exist, it may be difficult to obtain due to commercial confidentiality; data on defects and rework, for example, is not information that construction firms are keen to release if it has the potential to damage their reputation for quality. Methods are thus needed that produce at least reasonable results using data that is readily available.

Another point that should be considered is that the act of measurement may change the result. This notion is encapsulated in Goodhart's Law, which came originally from economics but was succinctly restated by Strathern (1997) as:

'When a measure becomes a target, it ceases to be a good measure.'

If management were to attempt to measure onsite productivity by, for example, observing bricklayers at work and counting how many bricks each laid in an hour or a day and the bricklayers are aware of this observation, then it is likely that they will work harder because they know that their performance is being measured. Common sense dictates that this will occur, and measurement must be done in such a way that results are not skewed; covert observation or observation over an extended period of time may be enough to fix the problem.

It is harder to deal with this phenomenon when considering broader measures of industry performance. Consider, for instance, time and cost overruns on construction projects – both outcomes are commonplace and may be perceived as indicators of poor performance on the part of the contractor. The UK government routinely collects such data as part of the Construction Industry Key Performance Indicators scheme. It could be in the interest of contractors to look for ways to report such overruns that would reduce the potential for failure to complete projects on time and within budget reflecting poorly on their reputation, and thus the act of measurement of these parameters has reduced their effectiveness as an indicator of performance.

Concluding remarks

This book addresses important aspects of an important subject, but there are aspects of construction measurement that are not addressed, for example, measurement of construction quality or the implications of the regulatory environment. The construction industry, construction researchers and construction policy makers need to recognize the complexity of construction measurement and insist on rigorous approaches to data collection, methods and presentation. Policy prescriptions are often based on research results, including data/measurement; poor data will often lead to poor conclusions and poor policy.

References and Further Reading

BCA (2012) *Pipeline or Pipe Dream? Securing Australia's Investment Future* (Business Council of Australia). www.bca.com.au/Content/101987.aspx

Best, R. (2012) International comparisons of cost and productivity in construction: a bad example. *Australasian Journal of Construction Economics and Building*, **12** (3), 82–88.

Bureau of Labor Statistics (2014) *Occupational Employment and Wages, May 2013, 47-2061 Construction Laborers* (United States Department of Labor). www.bls.gov/OES/current/oes472061.htm

Cordell (2014) *Commercial and Industrial Building Cost Guide*, Queensland. (Sydney: Reed Construction Data).

DLC (1999) *A Framework for International Construction Cost Comparisons* (UK: Davis Langdon Consultancy report for the Department of the Environment, Transport and the Regions).

Edkins, A. and Winch, G. (1999) *The Performance of the UK Construction Industry: An International Perspective.* Bartlett Research Papers (London: University College London).

Flanagan, R., Norman, G., Ireland, V. and Ormerod, R. (1986) *A Fresh Look at the UK & US Building Industries* (London: Building Employers Confederation).

Forrestal, L. and Dodson, L. (2102) High costs could kill big projects. *The Australian Financial Review*, 7 June. www.afr.com/p/national/high_costs_could_kill_big_projects_aujM8HwWPYsWPkwKh1t1DP

Freeman, I. (1981) *Comparative studies of the construction industries in Great Britain and North America: a review* (Garston: Building Research Establishment).

Hepworth, A. (2012) Local project costs 40pc above the US, says Business Council of Australia. *The Australian*, 7 June. www.theaustralian.com.au/national-affairs/local-project-costs-40pc-above-the-us/story-fn59niix-1226386836012

Langston, C. and Best, R. (2000) *International Construction Study* (Canberra: Department of Industry Science and Resources).

Langston, C. and de Valence, G. (1999) *International Cost of Construction Study – Stage 2: Evaluation and Analysis* (Canberra: Department of Industry Science and Resources).

Lynton (1993) *The UK Construction Challenge* (London: Lynton plc).

Meikle, J. (2011) *Note on Informal Construction*. International Comparison Program, 5th Technical Advisory Group Meeting, April 18–19 (Washington, DC). Available at http://siteresources.worldbank.org/ICPINT/Resources/270056–1255977007108/6483550–1257349667891/01.02_ICP-TAG04_Construction Note.pdf

Productivity Commission (2014) *Public Infrastructure Vols 1 and 2* (Australian Government). www.pc.gov.au/__data/assets/pdf_file/0003/137280/infrastructure-volume1.pdf; www.pc.gov.au/__data/assets/pdf_file/0005/137282/infrastructure-volume2.pdf

Rawlinsons (2014) *Australian Construction Handbook*, 32nd edition, Rawlinsons Construction Cost Consultants and Quantity Surveyors (Perth: Rawlhouse Publishers).

RSMeans (2014) *RSMeans Online* (Reed Construction Data). www.reedconstructiondata.com/rsmeans/rsmeans-online/

Strathern, M. (1997) Improving Ratings: Audit in the British University System. *European Review*, 5, 305–321.

Xiao, H. and Proverbs, D. (2002) The performance of contractors in Japan, the UK and the USA: a comparative evaluation of construction cost. *Construction Management and Economics*, 20, 425–435.

Editorial comment

Bill: "Look! We can have a really cheap holiday in Klajistan – the locals live on the equivalent of $2 a day!"

Bob: "How do they do it?"

While the people of Klajistan may not live in luxury, they can survive because the prices of food and other essentials in their country are low when expressed in Bill and Bob's national currency. Suppose the currency in Klajistan is the klaj and 1 dollar in Bill and Bob's currency buys 50 klaj when they arrive at the main airport in Klajistan – $2 then buys 100 klaj. If 1 kg of bread costs just 1 klaj, then $2 buys 100 kg of bread – suddenly $2 a day sounds like it could be enough to live on. The key factors here are exchange rates and price levels. If a Klajistani decided to holiday in Bill and Bob's home country (where bread costs $2/kg), he/she would find it a very expensive place, and a week's income would hardly keep them going for a day.

Differing price levels and fluctuations in exchange rates make it difficult to arrive at meaningful comparisons of cost between countries. Comparing costs in different currencies first requires that costs in local currencies be converted to a common currency. Money market exchange rates are seldom a good way to do this because they fluctuate in response to a range of factors such as changes in interest rates and investment patterns in different places as the money market's enthusiasm for different currencies changes depending on their profit-making potential for those who trade in them.

Choosing a way to convert national prices to a common base will depend on why the conversion is being done; for example, if a firm in Country A is considering funding the construction of a facility in Country B and that construction will be financed from cash reserves in the home country, then nominal (i.e. commercial) exchange rates do provide an appropriate way to do it, as the client firm will be transferring money in their home currency to Country B, and the amount in Country B currency will be determined by prevailing exchange rates. If, however, the intent is to make a theoretical comparison of the cost to build in different countries, then another method is needed. Exchange rates can move quite quickly while construction costs usually move relatively slowly. Thus the use of exchange rates in comparisons is likely to distort results, with comparative costs appearing to fluctuate in the same way that exchange rates can and do. Comparative costs could appear to vary considerably over quite short time periods due to changes in exchange rates, while construction costs may change little in the same time period.

This chapter outlines some of the theory that underpins a key concept, that of *purchasing power parity* (PPP). Using PPPs to express costs in various currencies in terms of a single base currency eliminates the effects of price-level differences and makes valid comparisons possible when the use of exchange rates is not appropriate.

2 Background to purchasing power parity indices

Göran Runeson

Introduction

Some 15 years ago, on a dark, hot and steamy night in Istanbul, I achieved a boyhood dream. I became a millionaire. I would like to say that it was the result of the kind of hard work or clever investments that my parents brought me up to believe in, or possibly even some ingenious crime. It wasn't. The truth is much simpler – I just changed a US $100 note into Turkish lira (TKY), which at the time was the least valued currency in the world. My $100 gave me well in excess of one million lira. A few years later, it would have given me almost five million.

However, any plans of a life in luxury on the French Riviera, rubbing shoulders with other rich people, would have been premature. Bread was selling in the streets for TKY10,000, and most things seemed to have four zeros too many attached to the price tag.

After a few days, when I got used to price tags that needed to be two or three times bigger than in Australia just to fit in all the extra print, we started to suspect that in general terms, it was cheaper, in Australian money, to live in Turkey than in Australia. It was not that everything was cheaper. Most things were, but many things were about the same price, and some things defied this common tendency by actually costing more. It was more that on balance, most prices seemed to be lower than in Australia, and some were a lot lower when converted into the same currency. That is commonly referred to as the price level in Turkey being lower than in Australia (MyTravelCost. com, 2013).

The price level is determined by the exchange rate. Had I received only TKY 0.5 million for my USD100, the price level in Turkey would have been a lot higher than in Australia, as in Australian money everything would have cost twice as much. The exchange rate, in turn, is influenced primarily by international trade and capital movements but also by a number of other economic and social variables. This book deals, in part, with how we can make more relevant and objective comparisons between standards of living or economic growth or of productivity in the midst of different price levels and exchange rates in different countries and what these differences actually mean. It also examines when it is legitimate to use the official exchange rate and when we need more sophisticated methods of measurement.

This idea of comparing the prices, and more directly the standard of living, in different countries has a long history. In the first modern text on economics, Adam Smith (1776, 1982) made comparisons between the standard of living of French and English workers and concluded, perfectly objectively, that while the English workers were very poor, they were not quite as poor as the French. With the limited and basic consumption of a preindustrial economy, such comparisons could be made fairly simply. How much food and shelter is it possible to buy on the average wage of a worker in each of the countries? Obviously those that could eat most and shelter best were, in a very real sense, better off.

With the development of more diversified and sophisticated economies, the complexity of this sort of comparison has been increasing. With the increased complexity, we have also seen an increasing interest in international comparisons of all kinds of economic variables such as standard of living, national product, productivity and economic growth, for all kinds of purposes, both from governments and their agencies and from individual researchers, investors, exporters and importers. This chapter is about the basic theoretical concepts behind various methods of comparisons and how they change depending on the purpose of the comparisons. In particular, it will concentrate on purchasing power parity (PPP) indices, methods that reflect real resource usage rather than monetary values modified through official exchange rates.

In the public sector, the interest in how to make comparisons is evidenced by, for instance, the International Comparison Program (ICP), a program that involves close to 200 countries. This program produces international price and volume measures for comparing GDPs and the various components of GDP, mostly using some form of specially developed PPPs (ICP 2011). The academic interest is evidenced from the current 25.4 per cent annual growth rate in the number of publications on the ways to best perform such comparisons (Clements *et al.* 2010).

In the public sector, there is extensive theoretical and empirical development of special-purpose indices by the OECD, Commonwealth of Independent States (CIS), ICP and the World Bank. In the private sector, we have a range of methods used for estimating relative price levels, starting from the somewhat frivolous Big Mac and the more recent but equally frivolous Starbucks Latte indices popularized by *The Economist* – which converts all prices to the number of Big Macs or cups of coffee we could buy instead – to special-purpose PPPs for goods or industries in two or more countries, compiled by various organizations.

There are good reasons why we construct special metrics like PPPs to compare prices and performances in different countries rather than using the official exchange rate. One is the volatility of exchange rates. In less than 10 years between October 2001 and August 2011, the Australian dollar

(AUD) went from buying USD0.51 to USD1.10 without any fundamental changes to the price level or productivity in either country (fxtop.com 2012). This increase is not evidence of an explosion in the standard of living in Australia. Rather, it is an indication that exchange rates are determined by a number of variables that have very little to do with the price level in either of the countries being compared. In this case, it was the investments caused by the resource boom and the comparatively high rate of interest in Australia.

Other reasons include differences in relative prices of products in different countries, differences in transport costs and taxes and barriers to trade – domestic and international – market conditions, resources and the like. As an example of price differentials, in July 2013, using the nominal exchange rate, the price of a litre of petrol in Venezuela was EUR0.02, and in Norway the price was EUR2.06, while the cost of a standard room in a five-star, centrally situated hotel was approximately 30 per cent higher in Caracas than in Oslo (MyTravelCost.com 2013). In real terms, this means that the visitor to Oslo gives up the equivalent of about 120 litres of petrol for a night in a good hotel, while his counterpart in Caracas has to give up the equivalent of about 15,000 litres.

Basic theory

The basic theory behind purchasing power parity is simple: if a good costs USD1.00 in the USA and AUD1.00 in Australia, the PPP Index (expressed as a ratio) is 1:1 using USA as a base. One USD buys the same in the United States as 1 AUD buys in Australia. If the prices are USD1.00 and AUD2.00, respectively, then the PPP Index is 1:2, or 1 USD buys the same in the United States as 2 AUD buys in Australia. The advantage with this PPP is that we get a metric that is totally independent of the exchange rates and all of the various factors that determine the level and variability of exchange rates.

The PPP depends on real goods and real resources only, and under the right conditions – it is argued – this PPP index can be extended from one good and two countries to any number of goods in any number of countries. When or if that happens, we have satisfied the conditions of the Law of One Price: the relative prices of goods in any one country are the same as in any other country, and the real exchange rate between any two countries, the PPP, is the same for any collection of goods.

So, under what conditions can we expect that a number of countries have the same real prices of goods and resources and so satisfy the Law of One Price? A partial answer can be derived from the so called Heckscher-Ohlin (H-O) trade theory for comparative advantage. In this, the most common reason for international trade, the H-O theorem – often illustrated as a 2×2×2 model as it uses two countries, two traded goods and two factors of production – sets out the minimum conditions under which trade will take place and what the effects will be on prices and costs in the respective markets. First, the H-O theorem demonstrates that all that is needed for trade to occur is a comparative advantage that is caused by, for instance, differences in tastes or different factor endowments. Second, it demonstrates that when trade occurs for this reason, it will have very specific impacts on the prices of goods and resources.

Factor endowment here refers to the proportion of the factors of production – labour and capital – in each country, and the theory proposes that in the country well endowed with labour but with little capital, labour will be relatively cheap and capital will be relatively expensive and vice versa. Technically, the marginal productivity of the abundant factor will be lower than that of the rare factor. In practice, this means that in the country with abundant labour, a good that uses labour extensively will be relatively cheap while the opposite applies to the country rich in capital. These differences in relative prices in the two countries are the precondition for trade. The country with abundant labour will have a comparative advantage in the production of the labour-intensive good and vice versa.

When there is an opportunity for trade, the labour-rich country will export the labour-intensive good and import the capital-intensive good. Expanding the production of the labour-intensive good will increase the demand for labour and therefore increase the price while the price of capital falls. For the trading partner, the production of the labour-intensive good will decrease in response to the import and demand for labour will decrease, and therefore the price will fall. Provided that a number of quite stringent conditions are satisfied, trade will continue until the price of the traded good and therefore also the price of labour is the same in both countries, and by that time, trade in the capital-intensive good has achieved the same thing with capital, and we have good and factor price equalization, that is, the relative prices of the traded goods as well as of labour and capital will be the same in both countries (Bhagwati *et al.* 1998).

The conditions that need to be satisfied are broadly the same as for a general equilibrium in economic theory: perfect competition, no impediment to trade, no transactions costs in addition to the specific condition that the available technology is the same in the two trading partners, which means that the goods will be produced in the same way in the two countries, and finally that factors of production cannot move between the trading countries. This means effectively that marginal and average costs will equal marginal and average revenue in all markets so that the respective industries in the two countries will be, to all intents and purposes, identical.

Taking this process one step further, we can intuitively see that the same thing is happening in the production of any nontraded good as well when there is international trade. Labour and capital will move into the sector with the highest reward until the price is the same in all sectors. In fact, looked at this way, trade is effectively just an alternative to moving the factors of production from one country to another.

The result of this, if we can extrapolate, is the so-called Law of One Price mentioned earlier. The production of any good will use up the same amount of resources in any country, and therefore the real price should be the same in all countries. The final instrument, the fine tuning, is *arbitrage*, which is when, if there are different prices in two markets, traders buy in the lower-price market and sell in the higher-price market, responding to demand and supply in a way that equalizes the prices. Whenever that is happening, we satisfy the Law of One Price, the condition for the use of PPPs (Persson 2008; Moffat 2012).

Obviously trade will occur also for reasons other than differences in factor endowment. We may, for instance, have trade when countries have absolute advantages in the production of goods due to access to specific resources, such as coal or iron ore or some agricultural products, or access to advanced technologies or venture capital or for a number of other reasons such as taxation or subsidies. Whenever that happens, there is no necessary move towards factor price equalization as we saw in the example of Venezuela and Norway, two countries with large resources of oil but different philosophies of how to use these resources.

There are various market distortions such as transport costs, trade barriers, regulations, taxes and subsidies that contribute to a failure of the international market. In such cases, the price equalization described by the H-O theorem will not occur.

The move towards globalization that we have seen over the last decades has also impacted trade. Globalization has, among other things, meant the removal of many of the impediments to trade such as tariffs and other restrictions and should therefore move trading partners towards factor price equalization. However, it has also removed many of the impediments to the international movement of capital. Capital-rich countries may now invest much more freely in countries with abundant labour, even for the production of goods they are not trading, and as a result maintain a higher return to capital in their own country and at the same time reap the rewards of a high return in the labour-rich country. This gives a country the potential to expropriate most of the gains of trade which, of course, ensures that factor price equalization does not occur (Chossudovsky 1997).

Rapid technological change and product development are also hampering any moves towards a stable equilibrium. Hence, while theoretically there should be a strong force towards factor price equalization, other forces are operating towards maintaining or even increasing current differences. The question now becomes empirical rather than theoretical: in reality, are the distortions discussed earlier substantial enough and widespread enough to dominate the equilibrating forces? And if so, does it matter?

There has not been a lot of direct testing, on a global scale, of factor price equalization. It seems that up to the end of the 1970s, inequality between countries increased substantially, while within countries it decreased. However, from the 1980s onwards, there are indications of a change towards more inequality within countries but less inequality between countries. This indicates either more rapid economic growth among developing countries, often, as in China, driven by international trade and/or more equal sharing of the benefits of trade due to moves towards factor price equalization, or a mixture of these effects (*The Economist* 2012).

Returning to the Law of One Price: if the law holds absolutely, the following equation would be true:

$$P_{Ait} = NE \times P_{Bit} = PPP \times P_{Bit}$$

where P_A and P_B stand for the prices in country A and B, respectively, in their local currency, for good i at time t, NE stands for nominal exchange rate and PPP for purchasing power parity.

This means that the price of good i at time t in country A is identical to the price in country B in the respective currencies of the two countries, adjusted for the nominal exchange rate, and that the purchasing power parity is identical to the nominal exchange rate. However, while possible, it would be much more likely, *a priori*, that the Law of One Price would apply in a relative way, revealing differences in price level between the countries being compared, so that

$$P_{Ait} = PPP \times P_{Bit} \neq NE \times P_{Bit}$$

where there is a difference between the nominal exchange rate and the PPP. This would be likely to happen whenever the countries were at different levels of economic development or for any of the reasons discussed previously.

However, for a PPP index to be useful in terms of measuring or comparing price levels, we would need to include a range of goods, preferably all goods traded within the countries. This means compiling an index covering a large number of goods, each of them weighted according to their importance in the index, which gives the price term for country A:

$$P_{Ait} = \sum_{i=1}^{N} w_i P_{Ait}$$

where the right-hand term is familiar from the calculation of conventional price indices.

If the purpose of the index were to compare real GDPs, the weights would be country specific and classified according to National Accounts. For comparisons of standard of living, the weights would also be country specific but include final consumption items only. In both cases, we are interested in the aggregate value of a bundle of goods, just as with GDP and per-capita consumption, only measured in terms of real resources instead of monetary values. For comparisons of productivity, there would be a common set of weights, as productivity is output per input but again measured in terms of real resources. By separating material which would be the same for the same product in different countries, and factors of production, which may not be the same, we can get an idea of productivity as we are measuring the real resources used.

For comparisons of growth rates over a period of time, we need to decide at which point in time the weights will be set: the beginning of the period, the end of the period or some sort of average over the whole period.[1] Each alternative has implications in terms of how the index reflects change. Laspeyres indices with the weights selected at the beginning of the period of comparison tend to overstate changes over the period of comparison. This is due to the substitution effect, as people would tend to change their consumption or production by substituting goods where the relative prices have increased for goods where the relative prices have fallen. The requirement for efficient consumption (production) is that the marginal utility (marginal productivity) of all goods divided by their price is the same. That maximizes total utility or total output. For the same reason, Paasche indices, with weights established at the end of the period, tend to underestimate cost increases. Deaton and Heston (2010) have calculated that the ratio Laspeyres:Paasche between United States and most European countries ranges between 1.05 and 1.10, but for extreme cases like Tajikistan it reaches 9.6. The commonly used pragmatic compromise, Fisher's index, with weights established according to their geometric mean, has problems with transitivity and so on. It is also true that for any form of comparison, the more similar the economies, in terms of factors such as tastes, income, resources and climate, the better any index will perform.

The Law of One Price and why it may not work

While in theory there is no difference between theory and practice, in practice there certainly is. As we have seen, we have a number of theoretical and practical reasons why the Law of One Price will not operate. They include transport costs, the consequences of any form of barrier against trade including quotas, and an example can be seen in the relative price differences in petrol versus accommodation in Venezuela and Norway described earlier. They also include anything that reduces the competition in one of the markets, including imperfect information. They also include the volatility of exchange rates mentioned, as that changes the relationship between the prices of imports and exports.

Purchasing power parity calculations are complicated when one or more of these reasons apply so that countries do not have a uniform price level but rather distinctly different price levels in different sectors of their economies. Also, for various reasons, people in different cultures and at different levels of income typically have different consumption patterns. That makes it necessary to convert the costs of different consumption or production patterns using a price index, and even at the best of times, this is a difficult problem. This is particularly so in instances in which goods widely consumed in one country are not available in others.

It may also be necessary to make adjustments for differences in the quality of goods and services. Hence, a PPP may change depending on which country is used as a basis or, as discussed, from what time the weights are taken. There are additional statistical difficulties with multilateral comparisons. Various ways of averaging bilateral PPPs can provide more stable multilateral comparisons but at the cost of distorting bilateral ones. These are all general issues of indexing; as Ravallion (2010) says, it is generally not possible to reduce complexity to a single number that is equally useful for all purposes.

The Balassa–Samuelson hypothesis

The theoretical problems that are associated with index numbers and exchange rates are well known and need not be restated here (for a discussion of how they apply to PPPs, see Ravallion 2010; Deaton 2010).

The problems associated with different price levels in different sectors of economies are described variously as the Balassa–Samuelson law or hypothesis, Harrod–Balassa–Samuelson effect (Kravis and Lipsey 1983) or Ricardo–Viner–Harrod–Balassa–Samuelson–Penn–Bhagwati effect (Samuelson 1994), and they refer to various aspects of the effects of differences in relative productivity or growth of productivity between the tradeable and nontradeable sectors of the economy.

It is outside the scope of this chapter to examine any of these effects in more detail. They are relevant here only in that they point to phenomena that are present in most countries and that make it difficult, both conceptually and operationally, to construct general PPP indices that incorporate both the tradeable and nontradeable sectors. The practical result is that we have, by and large, given up on general PPP indices for studies of specific industries, and we now prefer, whenever possible, to use special-purpose indices, including special-sector indices, instead of or in addition to the general purpose PPPs that are applicable to the whole economy. Specific-sector indices do not eliminate the problems, as there would be few products that do not require inputs from more than one sector but, pragmatically, they reduce them as much as possible.

Considerations for construction industry PPPs

As it is one of the major sectors in any economy, there have been numerous attempts to compare construction prices and productivities in different regions and countries, and several different approaches to the production of construction PPP indices have been developed. As emphasized earlier, PPP indices, like other types of indices, differ in terms of what they cover and how they measure inputs and prices. The following provides a brief overview of the most common ways of constructing PPP indices.

The simplest approach uses the output prices of one or more construction projects. These are standard projects using the prices paid by clients. While this approach is theoretically very straightforward, there are significant practical problems. For example, projects normally vary significantly from one country to another due to, among other things, climate, building regulations and level of economic development. In practice, it is difficult to obtain directly comparable construction output prices. It is most unusual for identical projects to be constructed in different countries at the same time in exactly the same circumstances. We normally end up with all the subjectivity that goes with estimated prices, as discussed later.

A somewhat more complex approach is to use input prices in the form of material and labour. Technically, in its simplest form, this approach potentially suffers the same problems as the previous in terms of differences in the physical aspects of the output, but it eliminates any differences in overheads and profits due to variations in the level of economic activity. An additional problem is that costs are frequently context dependent. They are, at least to some extent, dependent on the quantity used, the scale of the project, how and where items are used, who is doing the work, who is the customer for the work and what are the economic conditions at the time. A particular problem is that the relevant input prices are not purchaser prices, they are the prices paid by contractors to their suppliers, not the prices charged to their customers.

A third approach is to use intermediate or elemental prices of inputs. The prices can be based on contractor estimates or tender prices for parts of construction work. While suffering most of the problems of the previous approaches, it takes, to some extent, the guesswork out of determining the quantity of labour required for different types of work, although the calculations would still be *ex ante* rather than *ex post*. That is because the output prices typically used are not actual prices but estimates by consultants or contractors. While they may be based on extensive experience, they certainly contain room for subjectivity. Even where it is possible to calculate normalized average costs (e.g. cost per m²) for broadly similar types of buildings, the buildings may vary in quality and functional performance. At best they represent broad averages.

Where intermediate prices are available – price books or successful tenders – the ways they are used depend on the pricing approach of particular contractors and the specific context of particular projects. Price book prices or different tenders cannot reflect the specific circumstances of projects or the methods used by different contractors. Even if that were possible, prices and costs are known to vary considerably between similar projects in the same country or region. In economics, this is referred to as x-efficiency after an article titled "Allocative efficiency v X-efficiency" by Liebenstein (1966) in which he suggested, based on empirical studies, that there were often substantial differences in productivity between plants of similar designs in different locations. Such differences tended to persist over long periods despite market pressures. Later (1968, 1978), his studies of developing economies suggested that such inefficiencies tended to be greater in less developed countries due to more substantial market inefficiencies or failures as well as differences in factor qualities, adding a further complication to the construction of indices. Such differences in productivity would have a tendency to destroy or at least considerably modify any theoretical argument.

Finally, in addition to the basic approaches to PPPs outlined here, there are various combinations of the three approaches, which for obvious reasons combine the problems of the individual approaches.

Having determined the implications of different approaches, the next important issue concerns the weights of the different inputs. This is an explicit issue for the approaches that depend on input prices, but implicitly, it is also a problem where output prices are used directly. As mentioned before, it is a problem because buildings are not built the same way in different countries. Climate, tradition, building regulations and level of economic development ensure that it is, for instance, very difficult to compare rural residential buildings in Northern Europe and equatorial Africa.

However, there is also a problem with labour input. For traded goods, the productivity of labour, and for that matter, of capital, may be reasonably similar. However, both empirical evidence such as that presented by Liebenstein (1966, 1968 and 1978) and the Balassa–Samuelson hypothesis suggest that this is not likely to be true, especially for nontraded goods (Lipsey and Swedenborg 1996). Even fairly minor differences in weights may have dramatically different impacts. The problem is well illustrated in *The Economist's* (2004) calculations of PPPs using the Starbucks Tall Latte and the Big Mac indices. While neither of these indices is considered a proper PPP, they share their characteristics with PPPs in that they represent the prices of identical products – a Tall Latte or a Big Mac, respectively – in a wide range of different countries. This means that any price of any product in any of these countries can be expressed in a special "currency" – the equivalent number of Tall Lattes or Big Macs – simply by dividing the price of the good by the price of a Starbucks Tall Latte or a Big Mac.

The Economist's calculations showed that the indices derived from the Starbucks Tall Latte and the Big Mac are virtually identical for some countries like Canada (–16) or Turkey (+5 versus + 6), suggesting that the price level for the two products is approximately equal and that the nominal exchange rate is 16 per cent undervalued for Canada and, similarly, the calculated exchange rate is 5 to 6 per cent overvalued for Turkey. On the other hand, for China the indices are –1 and –56, respectively – this suggest that the nominal exchange rate is approximately right according to the

Tall Latte index but undervalued by 56 per cent according to the Big Mac index. For Hong Kong (+15 versus –45), the difference between the two indices is so large as to make use of the indices meaningless, yet both indices combine capital, property, machinery, labour, advertising, agricultural produce and other factors in combinations that are not too different. It is true that coffee is imported in many countries while the components of the Big Mac are mostly domestically produced, but on the other hand, coffee is not a major component in the price of a Starbucks Tall Latte (*The Economist* 2004). The answer is likely to depend more on tastes and marketing.

Approaches to construction PPPs

There are primarily three different approaches that have been used by international agencies over the years to collect price data for international construction price comparisons (ICP 2011). The basis for these approaches has been discussed already. The following section describes them briefly – for a more detailed discussion of these methods and of the most recent variant used in the 2011 pricing round of the ICP, refer to Chapter 4.

The CIS countries use a 'Method of Technical Resource Models'. Participating countries are required to collect data on wages and salaries in the construction industry, as well as cost data for building materials and energy products. Once collected, the input prices are entered into linear equations that are referred to as 'Technical Resource Models'. These models allow the pricing of 100 different residential buildings, nonresidential buildings and civil engineering works.

The second method is called the 'Standard Project Method' and is used by the OECD-Eurostat group. It is based on standard models for different types of construction projects. Participating countries are required to price each standard construction project by pricing a bill of quantities covering all the costs that constitute the purchaser's price.

Finally, the third method is the 'Basket of Construction Components'. It lies in between the input-based and the model-based approaches already described. It was developed and used for the 2005 ICP. Under this method, participating countries collected the direct costs of a basket of construction components. The direct costs include building materials, labour costs and hire of construction equipment. However, practical problems were encountered when the method was used in the field. In addition, the data obtained could not be readily aligned with standard national accounts frameworks as required for national income comparisons.

Following a thorough review of the various methods, a new approach was devised for the 2011 round of the ICP. The approach, developed under the framework of the research agenda of the Technical Advisory Group (TAG), is based on a selection of some 50 to 60 basic and common inputs to construction work weighted in each country to correspond with national construction practice (ICP 2011).

PPPs for measuring productivity

It is quite likely that either of these approaches will provide reasonable approximations of real construction prices. However, they will not provide answers to one of the most interesting issues of international construction comparisons, one that has occupied politicians, trade unionists and industry leaders alike: What about productivity? Is Australia as efficient as the United States? Or Hong Kong? The use of the official exchange rate or PPPs with country-specific weights for both labour and material effectively makes it impossible to make any inferences about productivity (Best 2012).

One possible way around this problem is to use constant weights for both labour and material. It does not account for multifactor productivity, but provided the materials are the same in all countries and are used in the same way, the difference between costs in different countries should be a measure of labour productivity, as in this approach labour would be the only factor to vary between countries.

Obviously, for all the reasons discussed, all materials will not be the same in all countries, so a slightly different approach is required as demonstrated by Best (2008). By creating and pricing a unit, which he referred to as a BLOC, consisting of materials and labour used across all countries in the proportions they would be used in an actual building, he derived a unit of international currency, the BLOC, that could be used as the Tall Latte or Big Mac of the construction industry. The materials used were things like concrete, glass, reinforcement steel or pipes, which are very similar across the world and together account for some 80 per cent of the total costs of materials of a particular type of building – in his case, a medium standard hotel. By calculating the costs of a similar hotel anywhere in the world in terms of BLOCs, he could bypass the official exchange rate and determine a real price. Say, for instance, that if a hotel in Sydney costs the equivalent of 50 BLOCs while a similar hotel in the United States costs 60 BLOCs, it shows that the hotel in the United States is more expensive in real terms. Since the material input is the same in each location, the cost differential must be due to more labour being used in the United States, or expressed in another way: US construction labour is less productive (Best 2010).

The strength of this method, the common weights in the different countries that make it possible to calculate relative labour productivity, comes at a cost – the approach is not as good for estimates of PPPs for the contribution to cost of living or the rate of growth as are the methods with country-specific weights. It is also likely to be less efficient when applied to countries with wide differences in technological development and/or with different rates of substitution of labour for capital.

Conclusion

There has, over the last few years, been a substantial interest in PPPs, primarily from international organizations, with attempts to establish a metric for comparing outputs and standards of living across different countries. PPPs offer a method of measurement that bypasses the problems with volatile nominal exchange rates and deals directly with real goods, services and factors of production.

PPPs are not without problems due primarily to deviations from the Law of One Price and problems associated with all forms of price indices. In particular, different price levels in the internationally traded and nontraded sectors have led to the development of special-sector PPPs, including those for construction.

Finally, it is suggested that while conventional PPPs are suitable, to various degrees, depending on market imperfections, transport costs, taxes and a variety of other issues, to compare national output or standard of living, changing the weighting of labour and material from country specific to uniform, makes it possible to construct a PPP that can be used to compare productivity in different countries.

Note

1 For good summaries of approaches to various aspects of aggregation, see United Nations Statistical Division (1992) or Deaton and Heston (2010).

References and Further Reading

Best, R. (2008) *Development and Testing of a Purchasing Power Parity Method for Comparing Construction Costs Internationally*. University of Technology, Sydney: Unpublished PhD thesis.

Best, R. (2010) Using purchasing power parity to assess construction productivity. *Australasian Journal of Construction Economics and Building*, 10 (4) 1–10.

Best, R. (2012) International comparisons of cost and productivity in construction: a bad example. *Australasian Journal of Construction Economics and Building*, 12 (3) 82–88.

Bhagwati, J., Panagariya, A. and Srinivasan, T.N. (1998) *Lectures on International Trade*, 2nd Edition (Cambridge: MIT Press).

Chossudovsky, M. (1997) *The Globalisation of Poverty: Impacts of IMF and World Bank Reforms* (London: Zed Books).

Clements, K., Lan, Y. and Seah, S.P. (2010) *The Big Mac Index Two Decades On: An Evaluation of Burgernomics*. Discussion Paper 10.14. University of Western Australia: Business School.

CNN/Money (2012) *Global gas prices 2012*. www.money.cnn.com/pf/features/lists/global_gasprices: Accessed 20 Feb 2012.

Deaton, A. (2010) Price indexes, inequality and the measurement of world poverty. *American Economic Review*, 100 (1) 5–34.

Deaton, A. and Heston, A. (2010) Understanding PPPs and PPP-based national accounting. *American Economic Journal: Macroeconomics*, 2 (4) 1–35.

The Economist (15 Jan 2004) The Starbucks Index: Burgers or Beans. *The Economist*. Available at: www.economist.com/node/2361072

The Economist (13–19 Oct 2012) For richer, for poorer: special report on world economy. *The Economist*, 1–28.

fxtop.com (2012) *Historical Rates*: http://fxtop.com/en/historates.php?C1= AUD&C2=USD&DD1=&M   M1=01&YYYY1=1990&B=1&P=&I=1&DD2= 23&MM2=01&YYYY2=2001&btnOK=Go%21: Accessed 24 Feb 2012.

ICP (2011) *International Comparison Program*: http://siteresources.worldbank.org/ ICPEXT/Resources/ICP_2011.html: Accessed 22 Feb 2012.

Kravis, I. and Lipsey, R. (1983) Towards an explanation of national price levels. *Princeton Studies in International Finance*, No. 52.

Liebenstein, H. (1966) Allocative efficiency v X-efficiency. *American Economic Review*, 56, June, 392–415.

Liebenstein, H. (1968) Entrepreneurship and development. *American Economic Review*, 58, May, 72–83.

Liebenstein, H. (1978) *General X-efficiency theory and economic development* (Oxford University Press).

Lipsey, R. and Swedenborg, B. (1996) The high cost of eating: Causes of international differences in food prices. *Review of Income and Wealth* 42 (2) (June), 181–294.

Lipsey, R. and Swedenborg, B. (2007) *Explaining Product Price Differences Across Countries*. NBER Working Paper Series, Vol. w13239. Available at SSRN: http://ssrn.com/abstract=999034: Accessed 20 Feb 2012.

Moffat, M. (2012) A Beginner's Guide to Purchasing Power Parity Theory (PPP Theory): Why Purchasing Power Parity Is Not Perfect: http://economics.about. com/cs/money/a/purchasingpower_2.htm: Accessed 24 Feb 2012.

MyTravelCost.com (2013) Price levels across 40 countries: Available at: www.mytravelcost.com/international-comparison/: Accessed 24 July 2013.

Persson, K. (2008) Law of one price. In: Whaples, R. (ed.) *EH.Net Encyclopedia*, http://eh.net/encyclopedia/the-law-of-one-price/: Accessed 24 Feb 2012.

Ravallion, M. (2010) Understanding PPPs and PPP–based national accounts: comment. *American Economic Journal: Macroeconomics*, 2 (4) 46–52.

Samuelson P.A. (1994) Facets of Balassa-Samuelson thirty years later. *Review of International Economics*, 2, 201–226. Accessed 1 Mar 2012.

Smith, A. (1776: 1982) *The Wealth of Nations* (Harmondsworth, Middlesex: Penguin Classics Reprint).

United Nations Statistical Division (1992) *Handbook of the International Comparison Programme Annex II Methods of Aggregation*, Series F No 62: http://unstats.un.org/unsd/methods/icp/ipc7_htm.htm: Accessed 3 Mar 2013.

Editorial comment

In 1599, Shakespeare gave one of his characters the line "Comparisons are odorous", almost certainly using it ironically knowing that it is was a misquotation of the phrase "Comparisons are odious". While comparisons may be odious, they are routinely carried out in industry, often as a way of assessing performance relative to others, typically in benchmarking exercises in which the performance of an operational unit (which could be an entire firm or some part of a firm such as an individual department or a regional office or some other business or management unit) is compared to that of another. The aim is usually to look for ways to improve performance once a benchmark has been set and best practice has been identified.

Comparisons of construction cost are a little different, although cost is often used as a single-factor method for comparing industry performance and even productivity, but how appropriate such uses are is debatable. Certainly construction cost comparisons have been attempted using a variety of methodologies for many years. In the preceding chapter, the notion of purchasing power parity (PPP) was introduced as a way of making costs in different currencies directly comparable, and this is now understood to be a key issue in international cost comparisons.

There have, however, been quite a number of comparative construction cost studies done, and there are ongoing exercises carried out by international firms of construction cost consultants, all of which involve, in some way, comparisons of the cost to build in different countries. Methods used to collect construction cost data and to bring costs in local currencies to a common base vary. Costs have been converted to a common currency using current, annual average and even 10-year average exchange rates; other studies have been designed to produce conversion factors (PPPs) to allow more meaningful cost conversions and thus more robust cost comparisons. Consultants collect and publish a range of international construction cost data that can be used to arrive at indicative costs to construct in different locations.

The International Comparison Program (ICP) is the most comprehensive of the studies, and construction forms only one part of the much larger exercise. Even within the ICP, the methods have changed over the years, and even now no single method is used throughout the whole of the ICP. This chapter describes some generic approaches to the collection and use of international construction cost data in comparative exercises and looks at a selection of past studies.

3 International construction cost comparisons

Rick Best and Jim Meikle

Introduction

Comparisons of the cost of construction across national boundaries have been routinely undertaken for many years for a variety of purposes. For clients, the reason may be as simple as wanting to know how much it will cost them in their own national currency to build a facility in another country; for governments and other agencies, it may be about how competitive their industry is in a global marketplace. For example, in 1949, a UK group, the Building Industry Productivity Team, visited the US (Anglo-American Council on Productivity 1950). A primary purpose of the visit was to compare the cost of construction in the two countries, which they did by comparing the cost of buildings of similar size used for similar purposes. The group concluded that US costs were in the order of 55 to 80% higher than comparable UK costs but noted that differences in design and specification of buildings varied considerably between the two countries. Costs were collected in the respective national currencies and then compared using the exchange rate current at the time of the group's visit to the US, GBP1 = USD4. A year later, that rate had changed considerably to GBP1 = USD2.80 – had this rate been used, US costs would have appeared to be more like 120 to 160% higher than UK costs. That there would be so much change in relative construction costs in such a short time is extremely unlikely.

Methods have not necessarily changed much over the last 60 years; Best (2012) analysed the methodology underlying claims made by the Business Council of Australia (BCA 2012) regarding the high cost of building in Australia relative to the cost in the US and showed that, using the same methodology with different currency conversion rates and data from alternative sources, building in Australia was arguably much cheaper than it was in the US.

While the terms *cost* and *price* are largely used interchangeably here, *cost* may be interpreted as the cost to the contractor to build, that is the cost of inputs, while *price* is the amount actually paid by the client to the contractor for the completed work. In the latter case, of course, the *price* paid is the *cost* to the client.

What sort of costs or prices are collected and how that is done often depends on the purpose for which the data is required, and there are a number of ways that it can be done. Cost data can be collected for whole projects, parts of projects or for inputs to projects, with each method/approach having its advantages and disadvantages. Significantly, the most recent round of the International Comparison Program collected data that was not for projects or parts of projects but for basic inputs – this is described in the following chapter. The examples given highlight a further issue that arises when comparing costs between countries, which is how costs in different currencies can be 'normalized' – converted to a common currency unit – and thus be directly compared. Both aspects of cost comparisons are discussed in this chapter.

Collecting cost data

Whatever the purpose, a number of approaches are used, and have been used over the years, to collect cost data for international construction cost or price comparisons. These have been variously based on:

- output prices of construction (the prices of completed construction projects charged by contractors to their customers)
- input prices to construction (the prices paid by construction contractors for the inputs to construction work – primarily construction materials and labour)
- intermediate prices (prices used, for example, by contractors for parts of construction work, either in their own estimates or in tenders to customers)
- some combination of these.

Cost data may be based on complete projects or on some combination of elements that can be taken to represent whole industries or 'all construction'. Where comparisons are made at a whole-of-industry level, they may involve some sort of weighting that reflects the share of total construction expenditure that each component represents, or they may be unweighted. The weighted versus unweighted approach is discussed in the following chapter in the context of the two most recent rounds of the International Comparison Program (ICP), where the 2005 round had a three-tier system of weights and the 2011 round features a combination of weighted (for industry sectors and resource groups) and unweighted (for individual resource inputs) approach.

Ideally, the prices collected should be output or purchaser prices, the actual final price paid by clients for completed work. Such prices include the cost of all inputs plus the other costs that make up the final project cost, including profit, taxes, contractors' overheads and professional fees paid to designers and other consultants. They also reflect both productivity and labour/equipment use, which will vary between locations.

A number of international consultancies regularly collect and publish superficial (i.e. cost per m²) building costs, usually expressed as cost per m² of floor space. Costs are given in the currency of each country, often converted to some single currency (typically USD, GBP or EUR). The use of cost/m² data appears, *prima facie*, to offer a relatively simple method for comparing costs between locations. The basic notion is that if it costs USD1000/m² to build a certain type of building in the US and AUD1500/m² to build the same type of building in Australia, then the two amounts in the different currencies buy the same volume of construction. In reality it is not so simple. The key concerns are availability of data, the consistency of data between countries and whether the data represents national average prices for buildings that are actually similar enough to be compared.

Data is available for many countries but is often not readily available for countries other than those included in the data published by the major consultancies. Where data can be obtained, it is often unclear how reliable the data is in terms of consistency of approach to factors such as the measurement of gross floor area (e.g. does it include unenclosed covered areas or not), and the specifics of the building projects, such as what is included or excluded (e.g. fitout to office space), are not always provided, and thus the comparability of the projects is unknown. Equally, it is seldom clear whether the projects for which costs are given represent 'typical' construction in each location.

Statisticians tend to use terms like 'goods', 'products' and 'services' for items such as televisions, cars and haircuts for which costs can be collected. Construction projects vary considerably in size, type and complexity and typically have no typology by which they can be separated into anything other than broad functional groups such as hospitals or offices; as a result, they seldom fit well into the statisticians' categories and are thus harder to price.

Comparability and representativeness

Finding construction that is both similar enough to be readily compared between countries and that represents typical construction in each location is not a simple task, and the problem is not limited to the construction sector. Even within a country, something as simple as a haircut varies considerably in cost and quality between, say, a small shopfront in a rural town and an up-market salon in a major city. Construction projects vary even more markedly, particularly when international comparisons are the aim, due to design for different climatic and seismic conditions, differences in regulations and client expectations. It is a main reason construction has been described as 'comparison-resistant' by the World Bank (Walsh and Sawhney 2004). While they may exhibit many similarities, most building projects are unique, even within a single country, when considered at a detailed level.

Differences between countries mean the nearest we can get to identical projects are generally functionally similar projects such as basic warehouses, while more complex projects such as hospitals, bridges and schools vary considerably and pose larger problems.

Attempts have been made to look at smaller parts of projects and basic inputs such as materials and labour in order to achieve greater comparability and representativeness; these are discussed in this chapter.

Collecting output prices

With methods based on output prices, the preference is to collect actual or out-turn prices, that is final prices paid by clients for building work. In practice, it is difficult to obtain such data. It is most unusual for identical projects to be constructed in different countries at the same time in exactly the same circumstances. The output prices that are typically used are therefore either consultants' or contractors' estimates, and thus they are not 'real' prices; while they are based on experience and represent an expert view on a range of projects, they must contain an element of subjectivity. Even though it is usually possible to obtain average real prices (e.g. cost per m² of floor area) for broadly similar project types (e.g. mass-market housing, industrial buildings, office buildings or sections of highway), these are not identical and are likely to vary in quality and functional performance. This is not necessarily a problem, as clients planning to build outside their own country will build to meet local standards, and thus it is local costs that are relevant to them. If, however, the aim is some sort of more general comparison of industry performance or international competitiveness, then differences in project design, building codes and typical materials can distort the outcomes.

Standard projects

As it is virtually impossible to find identical buildings in each location that can be compared, this method is generally based on the pricing of one or more hypothetical building projects; if more than one project is included, usually a range of different building types are priced. Project costs are determined by the pricing of detailed bills of quantities (BQs), in which each project is broken down into a number of detailed work items that are individually priced by construction cost experts in each location. If comparisons are being made at the whole-of-industry level, then project costs are aggregated, perhaps with the various projects weighted according to the volume of each type of building constructed in each location. The standard projects method is the method currently used by the OECD-Eurostat program to produce construction-specific purchasing power parities (CPPPs) as part of a broader exercise that produces GDP–level PPPs.

This approach:

- is designed to collect prices that are best estimates of actual prices paid; however, it is typically tender prices that are collected, and these are in reality forecasts of the cost to build and are thus not 'current prices' as is the case with other PPP price data.
- should reflect capital/labour mix and relative productivity differences.
- requires projects that are both representative and comparable across all locations, which is seldom possible.
- is designed to collect data for different construction types (residential, nonresidential and civil engineering) and 'all construction' by using a range of project/building types.
- is difficult to adapt to cover work other than the construction of new projects.
- can be costly in implementation, particularly if more than one price observation is collected for each project from each location, although if a construction cost expert is employed, then their experience should allow them to provide a more considered and balanced set of prices than might be the case with an observation from a single-industry participant such as an estimator in a contracting firm.

Project prices

This method is based on what are sometimes referred to as *area cost models* or *matched models*. Historical cost data in the form of rates per m² of floor area ($/m²) are collected for various building types, and these are used as an indicator of relative costs across countries by a number of firms of construction cost consultants. It is a relatively crude method, however, as costs are generally expressed as ranges that reflect the great variety of designs and construction methods that may occur in any category of building – for example, an 'industrial building' may take many different forms even within a single country.

Main features:

- based on actual output prices
- relies on databases of historical data, which may or may not exist in all locations
- relies on consistent data collection and analysis
- based on average prices that may or may not be truly indicative of actual prices (i.e. depends heavily on the availability of representative data for various typical building types, and those buildings need to be comparable and representative)
- assumes use of common concepts for measurement and content.

Table 3.1 shows a sample of such data published by an international consultancy. Costs of this nature are generally given by city rather than

Table 3.1 Cost/m^2 data for three styles of hotel published by an international consultancy (Source: Davis Langdon 2011)

Hotel	Sydney	Auckland	Bahrain	Hong Kong	Bangkok	Los Angeles	London
3-star budget	2060	1865	1920	2440	1275	2100	1485–1845
5-star luxury	3160	2420	2700	3105	1800	4500	2570–3530
Resort style	2810	1860	3300	N/A	2110	4500	N/A

as national averages, particularly for countries such as the US where there are significant regional variations. In some cases, even in a single location, costs can vary considerably, as illustrated by the London rates shown in the table, where the higher rate for three-star budget hotels is around 25% higher than the lower rate, while for five-star luxury hotels, the difference is nearly 40%.

Using input prices

In this method, a selected set or 'basket' of inputs or resources is priced. Inputs include typical construction materials (possibly including some components, such as windows, that are typically manufactured off site), various types or classes of labour and possibly some items of construction plant and equipment. Basket costs are adjusted to include estimates of site-specific and general overheads and builders' profit. Other adjustments may include local taxes and consultants' fees.

Comparable input prices are relatively straightforward to collect. They are for standard measured units (e.g. tonnes, m^2, m^3, days) of readily available items (e.g. concrete, timber, steel, skilled labour). A particular issue with this approach is that input prices are not purchaser prices, they are the prices paid by contractors to their suppliers, not the prices charged to their customers. Furthermore, input prices (and, indeed, all construction prices) are context dependent, that is they are, at least to some extent, dependent on the quantity used, the scale of the project, how and where it is used, who is doing the work, and who is the customer for the work; prices collected thus have to be some sort of average that accounts for these variations.

Main features:

- generally reasonably easy to implement
- in line with many other price index and price comparison methodologies
- provides current price levels rather than forecasts (which is, as noted earlier, what tender prices are – out-turn costs would be preferable, but they are seldom available, so current 'real' prices for inputs may be the next best option)

- reduced effort offers the possibility of multiple observations and/or more frequent surveys and makes validation of data by reference to data providers feasible
- designed to provide price levels for different construction types (residential, nonresidential and civil engineering) and 'all construction' by using different baskets of resources
- does not implicitly account for contractors' margins (provides contractors' input prices rather than output prices) or other general cost inputs, so these must be estimated or collected and added separately or possibly ignored.

Using intermediate prices

Intermediate prices are composite rates or prices for items of work and, as such, combine the cost of all required inputs (materials, labour and so on). Such prices are available – from price books and successful tenders, for example – but the way they are used depends on the pricing approach of particular contractors and the circumstances of particular projects. They need to be used consistently; for example, if prices from successful tenders are used, all prices need to come from the same tender, and even then there is no guarantee that margins have been evenly distributed across the items in the tender, as tenderers may weight some items to enhance their cash flow. Price book prices cannot reflect the specific context of projects or the methods used by different contractors. Due to these factors, it is likely to require expert input to price the items in any basket. Another challenge is to find items of output that are sufficiently similar in all locations that are both representative and comparable.

The premise for this method is that by pricing composite yet discrete items of construction output, not only does the data represent purchaser prices, but these prices inherently reflect differences in productivity and labour/equipment ratios between locations. This component cost method is similar to the BQ approach; however, rather than pricing every part of a project, a standardized set of components is taken to represent whole projects or even a whole sector; this greatly reduces and potentially simplifies the price collection process. Typical components include concrete footings, an area of finish (e.g. cement render) to a surface, a concrete pier, or an area of pavement. The items are typically tightly specified with precise details of size, materials and construction method.

Main features:

- designed to reflect productivity and labour/capital differences between locations
- includes contractors' margins (if correctly priced)
- designed to provide price levels for different construction types (residential, nonresidential and civil engineering) and 'all construction'

- difficult to specify components that are sufficiently similar in all locations and reasonably representative of every location.

Examples of implementation of various methods

Each of the generic methods described has been implemented at least once. Several methods provide data for the ICP, and its component parts are discussed further in the following chapter. Other examples of the use of methods and some variations to the basic methods are discussed here. The following sections give overviews of a number of construction cost studies carried out over the past 25 years. Some are stand-alone studies, while others have been, and in some cases still are, conducted regularly by commercial organizations. They illustrate similarities and differences in approach and often demonstrate the generic problems that are faced by those carrying out such studies.

UK Building Employers Confederation – 1986

Construction costs were a key part of an investigation carried out in 1986 that compared many aspects of the construction industries in the UK and US (Flanagan *et al.* 1986). Part of the exercise involved direct comparisons of construction cost between pairs of completed projects that covered a range of building types, including residential, office and industrial. A set of nine suitable projects was identified in the US and an extensive search was then conducted to find similar UK projects that matched each of the US projects as closely as possible. Although every effort was made to match the pairs, basic differences in a number of factors including quality, regulatory requirements and client expectations meant that the cost data had to be adjusted by the researchers in order to improve the comparability of the projects. This is a fundamental issue and one that is not easily resolved, as any such adjustment is likely to require some degree of judgement and subjectivity, yet it is likely to be necessary if valid comparisons are to be drawn. Identifying suitable projects in this way is a lengthy process even for a small sample of nine pairs, and the smallness of the sample raises questions about how representative of all construction the selected projects could be, particularly in the US context given the substantial differences in construction costs that exist between different parts of the US. As the researchers acknowledged, regional differences in cost are also apparent in the UK, but they are of much greater significance in the US.

The method used for comparing costs employed a sort of construction-specific purchasing power parity (CPPP), as the project costs were expressed in national currency units per m^2 of floor area and the cost ratio for each of the pairs calculated (e.g. for one pair, the rates were GBP288.00/m^2 and USD502.90/m^2, respectively, with a ratio of 1.75). Five of those ratios (range: 0.86–1.02) were below the then-current exchange rate and four were near or above (range: 1.22–2.55), thus suggesting that the exchange rate was a reasonable proxy for a purchasing power parity at that time. In fact, if the geometric means of the nine GBP and USD rates are computed and compared, a CPPP (GBP:USD) of 1.21 results, and that is reasonably close to the average exchange rate for 1985 of GBP1 = USD1.30 (fxtop, 2013). The original results were a little clouded, however, as costs for each pair were adjusted to bring costs for each pair to a common base date, but all the pairs were not brought to the same base date; thus the pair-wise comparisons appear valid, but it is hard to judge the robustness of the overall comparison.

Davis Langdon – standard projects

Davis Langdon Consultancy (DLC 1999) used a set of three standard building projects, comprising a single-storey factory/office (GFA 2,900m²); a multistorey office block (GFA 4,500m²) and a single-family dwelling priced in the UK, Ireland and the Netherlands to compare construction costs among three locations.

The Davis Langdon study produced bilateral comparisons that separated cost differences due to differences in specification as well as those due to differing price levels. The basic methodology is shown diagrammatically in Fig. 3.1. A typical bilateral comparison is shown in Fig. 3.2. The buildings were priced by experts in each country in national currencies. Total costs were then converted to European Currency Units (ECUs were an informal predecessor to the euro) to allow direct comparisons of cost.

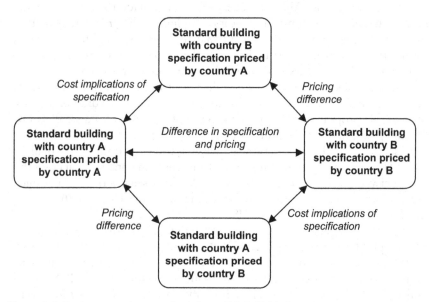

Figure 3.1 Basic comparison methodology (DLC 1999:4)

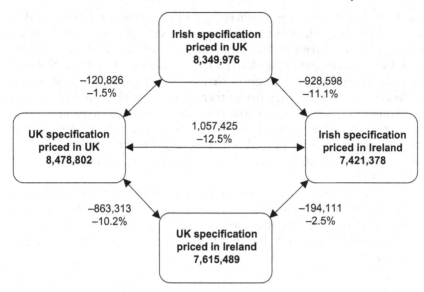

Figure 3.2 Office building price comparison – project prices in ECUs (base date September 1997) (DLC 1999:18)

The method was expanded to include differences in building geometry for residential buildings that were typical of the different locations. This allowed analysis of overall cost differences between locations to show the relative effects of the three variables, viz. price-level differences, specification differences and differences in building geometry (form).

Figure 3.2 shows that the 'standard' UK building is 12.5% higher in price level than the 'standard' Irish building and that that is largely (10.2–11.1%) made up of differences in price levels and only a little (1.5–2.5%) by differences in specification.

Davis Langdon – basket of materials and labour

In 2003, Davis Langdon undertook a test exercise of a basket-of-goods approach to international construction cost comparisons. The basket consisted of 35 materials or products and four items of labour. Prices were sought for large, medium and small quantities/projects (indicative quantities were provided for each); in practice, only one set of quantities, that for small projects, was consistently priced, and the test exercise was eventually based only on that data. Weightings for the basket were derived from the 1999 UK Input-Output (I/O) tables, with one or more items being taken as proxies for either industries (materials or products) or compensation of employees (labour). Value weightings were converted to unit quantities using the UK prices.

The weighting approach used was unusual and probably unique; rather than producing a standard weighted basket and then seeking price data for that basket, unit rates were gathered, and these were used to assemble a basket of goods and labour that contained the same list of materials, products and labour categories but different quantities for each item. Quantities for each country were derived from national I/O tables. National accounts generally include figures that represent purchases of materials by the construction industry from other sectors, and these provide a value weighting of inputs to the construction industry. Priced items in the basket were assigned to various product groups in the I/O tables and the average contribution to the basket determined. The labour:materials ratio was assumed to be 40:60 with no separation of plant costs or profit and overheads.

This exercise was not taken any further at the time, but it was revised and reviewed in the methodology development stage for the 2011 ICP. Its main general shortcoming is that it can usually only be used for 'all construction'; the other problem for the ICP is that I/O tables are not available for many countries.

A standard projects example

In 1999, a two-stage comparison study was carried out for the Australian government that compared Australian construction costs against costs from six other countries. In the first stage, seven hypothetical projects of various types were priced in local currencies in the seven target countries. Project costs were converted to AUD using current exchange rates. Based on those figures, the authors of the first-stage report suggested that "it would appear that Australia *with its currency exchange advantage* is more than competitive with the other Western countries and on a par with Singapore" (Page Kirkland 1999:2 – emphasis added).

In Stage Two of the study, Langston and de Valence (1999) acknowledged the problems of using exchange rates to convert costs. They noted, for example, the effects of fluctuations in exchange rates between Australia and some of the other countries around the time of the study, which produced very different results if varying exchange rates were used over the course of just one year.

In an effort to eliminate the problem, they utilized the Big Mac Index (for an explanation, see *The Economist* 2004) and expressed costs in local currencies as the number of hamburgers that (at local prices) equated to building cost. On that basis, Australia appeared to be more expensive than all but one of the other six countries (China) included in the study and the most expensive of the four Western countries, that is, the cost to build equalled the cost of a greater number of hamburgers in Australia than it did in the other countries. Subsequent work by Langston and Best (2005) did not provide strong evidence to support the hamburger exchange rate in the construction context, but the concept of purchasing power as a method for bringing costs to a common base has been explored further (e.g. Best 2008).

The BLOC method

In early 2008, a comparative construction cost study using a basket of goods and labour was carried out across six locations: three cities in Australia plus Auckland, Singapore and Phoenix. The basket was dubbed a BLOC, a Basket of Locally Obtained Commodities (Best 2008).

A bill of quantities for a completed hotel, priced by the successful tenderer, was analysed to identify the most cost significant materials in the most cost significant trade or work sections. These materials were then taken to be representative of all the materials in each trade. Total values for the various trades were used as the basis for deriving weights (quantities) of the various materials. Composite unit rates for the significant items were broken down into materials, labour and plant or equipment costs, and published labour rates were used to determine the quantity (hours) of various classes of site labour required to install the derived quantities of each material.

A survey instrument comprising the fixed-weight basket of materials and labour supplemented by several general questions regarding site-specific overheads, general overheads and profit margins was priced by estimators and quantity surveyors in the six target locations. Average unit rates were calculated for each item for each location, and these were used to calculate the average cost of the BLOC in each location. At this stage, BLOC costs represented input costs only. Margins were added for site-specific overheads (referred to as Preliminaries in the UK and Australia) based on estimates from the survey respondents, and similar adjustments were made for general overheads and profit. Finally, the appropriate local taxes were added to produce BLOC costs that represented purchaser or output prices rather than input costs. Using Phoenix as a base, city-based construction-specific purchasing power parities (CPPPs) were calculated by dividing the cost of the BLOC in each location by the cost of the BLOC in Phoenix.

While the basic method appears sound, the exercise only provided price levels for a single building type (a hotel), which may or may not be an appropriate proxy for all nonresidential construction, let alone for all construction. Further application of the method using a range of building types may show whether the single building type is a suitable proxy. Another difficulty is that the data necessary to convert input prices to output prices (e.g. overheads and profit) is only an estimate and is hard to validate.

BCIS: Asia Building Construction Survey

In August 2006, the Building Cost Information Service (BCIS) published the first issue of the *Asia Building Construction Survey*. The report was survey based, and amongst the data gathered was average costs per m² of gross internal floor area for 10 building types as assessed by respondents in the various countries. These were reported in local currencies, with costs converted from local currencies into USD, EUR and GBP using current exchange rates.

The aim of the study was to produce a commercial document that would provide companies wishing to build facilities outside their home country with indicative costs for such facilities compared to the cost of building in their home country. It included costs for a range of materials and labour inputs, but no attempt was made to aggregate these in any way.

Cost models for engineering construction

The International Cost Engineering Council (ICEC) launched a project in 2008 with the aim of producing location factors to assist in early estimates for projects located in different places. Two cost models based on a handful of composite items were to be priced by ICEC members in different countries, but as participation was purely voluntary, little or no data was ever collected. The cost models were for a building and a process plant with most of the composite items that were priced being common to both models/projects. Respondents were asked to provide composite unit rates including labour, materials and plant costs for around 10 components (e.g. an *in situ* reinforced concrete column, a plasterboard and metal stud partition wall). Additional data including hourly rates for various categories of labour was to be provided by respondents. The survey form and the accompanying pricing notes appear to be quite US centric, and although the survey was brief (in terms of the number of items to be priced), actually completing the survey would have been quite challenging. Weights (i.e. quantities of each item) in the models reflect the cost engineering basis of the survey: the building model, for example, includes just 50m² each of external wall and internal wall but 500m of 12.5mm-diameter copper pipe and 5 tonnes of structural steel.

International cost data published by consultants

Indicative costs are routinely collected and published by a number of international firms, with some data freely available on websites and other data only provided to paying clients. Firms that are active in this area include:

- Rider Levitt Bucknall
- Gardiner & Theobald
- EC Harris
- Turner & Townsend
- Davis Langdon/AECOM
- Faithful+Gould
- Rawlinsons
- Reed Construction Data.

Several exercises are outlined later in this chapter. It is important to note that these datasets are often collected and published more for publicity than for application. They normally carry disclaimers that point out that the published data should only be used to give a general indication of relative costs.

EC Harris: International Building Costs

EC Harris is an international 'built asset consultancy'. They publish an annual survey of building costs across around 50 countries with cost/m² rates for a broad range of building types (EC Harris 2012). Costs are collected in national currencies and converted to USD, GBP and EUR as necessary. They acknowledge that relative price differences can be substantially affected by fluctuations in exchange rates as well as factors such as location, accessibility and design intent. As a result, the cost data has limited application apart from providing a snapshot of comparative costs at a given time and in circumstances where exchange rate conversions are appropriate.

Gardiner & Theobald: International Construction Cost Survey

This international firm of construction cost consultants regularly publishes the results of an *International Construction Cost Survey* that includes cost/m² rates for eight building types and rates for 11 materials and three classes of labour across a range of countries around the world (Gardiner & Theobald 2011). The published data gives no indication of any quantities/weights for the input items; however, it can be assumed that the aim of the exercise is not to produce a cost for a basket but to indicate the comparative cost of individual materials and various categories of labour between countries. Costs are published in USD, EUR and GBP, with costs in national currencies converted to each of the base currencies using current exchange rates.

Turner & Townsend

Another global firm, Turner & Townsend, is a 'program management and construction consultancy' with offices in more than 30 countries. They produce an annual *International Construction Cost Survey* (Turner & Townsend 2012) that covers not only cost/m² rates for a range of building types but also detailed information on labour and materials costs as well as rates for a number of complete work items such as an area of soffit formwork and an area of wall tiling. Costs are published in national currencies with an exchange rate based conversion to USD.

Commentary

The examples described in the preceding sections do not represent an exhaustive review of every attempt made to collect and compare international construction costs, but they illustrate an ongoing interest in such exercises covering more than 60 years. In later chapters, some more recent exercises are considered, including the 2011 round of the ICP, which is, at the time of writing, the most recent pricing round for that program.

A number of common threads can be identified across the various studies even though the purpose of the various types may vary. The costs collected generally fall into three categories: most commonly whole-project costs expressed as cost/m² of floor space and/or costs for basic inputs that usually comprise labour and materials/components, with some instances of intermediate prices for items of construction work.

The building types for which area rates are collected are generally those that can be considered to be quite generic: schools, offices, warehouses and so on; however, the scope for variations between individual projects both within and between countries is large, and cost data can be affected by many factors that make direct comparisons hard to validate. Apart from design and specification differences, there is little consistency in the way 'floor area' is measured between countries. This is well illustrated in a study conducted by the Council of European Construction Economists (CEEC 2004). Table 3.2 shows the differences in gross floor area (GFA) for an identical building as measured using local measurement conventions in a number of European countries.

Table 3.2 GFA for a typical building as measured in various European countries (UK base) (CEEC 2004)

Country	GFA (m2)	Variance
UK	2585	—
Switzerland	2875	+11%
Holland	3007	+16%
France	3412	+32%
Finland	2758	+7%
Denmark/Spain	1800	−30%

The markedly lower GFA reported in Denmark and Spain is a result of basement floors being excluded from GFA in those countries. The remaining countries, however, show larger GFAs than the UK, and while the CEEC does not provide any explanation for these variations, it is likely that they relate to factors such as the inclusion or exclusion of areas such as voids at lift and stairwells and floor area occupied by internal and external walls. Such measurement may be carried out reasonably consistently within a country, but there is certainly no standard set of rules that is applied across national boundaries.

Direct comparisons of cost data published by different firms can show large variations in cost/m² rates for what appear to be similar building types. Best (2012) demonstrated such variation where data from several equally authoritative sources produced very different results when fed into a single analysis, for example where data from one source showed hospitals in Australia costing 62% *more* than hospitals in the US, using the same currency conversion method, but data from a similar source showed Australian hospitals costing 45% *less* than similar buildings in the US.

In order to compare costs across countries, it is generally necessary to convert costs to some common base currency. While money market exchange rates are routinely used for this purpose, that method is only appropriate where the aim of the comparison is specifically to arrive at how much of one national currency would be needed in order to finance a construction project in another country with a different national currency at a particular time. The method is, therefore, not appropriate if the aim is to make a theoretical comparison of construction costs in terms of what can be called opportunity costs, or how much consumption of goods and services would be foregone in order to direct spending to construction. The Big Mac Index mentioned earlier illustrates the idea of opportunity cost and purchasing power parity: if the cost of a project in a local currency is divided by the local cost of a standard hamburger, the project cost is then expressed in a common 'currency', the hamburger. When the cost of the same project in the local currency of another country is similarly divided by the cost of a hamburger in the currency of that other country, then again the project cost is expressed in the 'hamburger currency' and the two project costs can be directly compared. If one cost is equivalent to fewer hamburgers than the other, then at least in terms of hamburgers, the cost to construct in that country is lower than it is in the other, and the lower cost to construct means that the person or entity intending to build will sacrifice less opportunity to buy other things (such as hamburgers) in order to spend their money on building.

Whether the Big Mac hamburger is a suitable unit of 'currency' for expressing building costs is open for debate, but the theory behind the Big Mac Index, that of purchasing power parity (PPP – see Chapter 2), is now generally accepted as a means for eliminating or at least reducing differences in price levels between countries. The ICP, which is explored in the following chapter, uses PPPs across a number of economic sectors, including construction, to make comparisons that include GDP per capita that do not involve costs that are brought to a common base using exchange rates. While studies such as the annual Turner & Townsend survey contain a lot of useful base information, there has typically been no attempt made to eliminate price-level differences; however, Turner & Townsend's 2013 international cost report includes, for the first time, purchasing power parities (Turner & Townsend 2013). If sound comparisons of costs between countries are to be made, then the cost conversion issue must be addressed, as regardless of the amount or quality of the data collected, the results of comparisons will have limited application unless the cost data is handled in such a way that the effects of fluctuations in nominal exchange rates are removed.

References and Further Reading

BCA (2012) *Pipeline or Pipe Dream? Securing Australia's Investment Future* (Business Council of Australia). www.bca.com.au/Content/101987.aspx

BCIS (2006) *Asia Building Construction Survey* (London: Building Cost Information Service).

Best, R. (2008) *Development and Testing of a Purchasing Power Parity Method for Comparing Construction Costs Internationally.* University of Technology, Sydney: Unpublished PhD thesis. Available at: http://works.bepress.com/rick_best/

Best, R. (2012) International comparisons of cost and productivity in construction: a bad example. *Australasian Journal of Construction Economics and Building,* 12 (3) 82–88.

CEEC (2004) *The CEEC Code of Measurement for Cost Planning* (Council of European Construction Economists). www.ceecorg.eu/eurupean-code

Davis Langdon (2011) *The Blue Book 2011* (Australia: Davis Langdon).

DLC (1999) *A Framework for International Construction Cost Comparisons.* (London: Davis Langdon Consultancy report for the Department of the Environment, Transport and the Regions).

DLC (2003) *An Initial Test Exercise on a Basket of Goods Approach to Construction Price Comparisons.* Unpublished (London: Davis Langdon Consultancy).

EC Harris (2012) *International Building Costs* (London: EC Harris).

The Economist (15 Jan 2004) Burgers or Beans. *The Economist.* Available at: www.economist.com/node/2361072

Flanagan, R., Norman, G., Ireland, V. and Ormerod, R. (1986) *A Fresh Look at the UK & US Building Industries* (London: Building Employers Confederation).

fxtop (2013) *Major Historical Exchange Rates* http://fxtop.com/en/historical-exchange-rates.php

Gardiner & Theobald (2011) *International Construction Cost Survey.* www.gardiner.com/assets/files/files/a47a4dcd5d29f0071e00bf2d6c589851ed552cc7/ICCS%202011%20$%20Version.pdf

Langston, C. and Best, R. (2005) Using the Big Mac Index for comparing construction costs internationally. In: *Proceedings of COBRA/AUBEA Conference.* Queensland University of Technology, July.

Langston, C. and de Valence, G. (1999) *International Cost of Construction Study – Stage 2: Evaluation and Analysis* (Canberra: Department of Industry Science and Resources).

Page Kirkland (1999) *International Cost of Construction Study. Stage One – Base Cost of Construction.* (Canberra: Page Kirkland Partnership study for Department of Industry Science and Resources).

Turner & Townsend (2012) *International Construction Cost Survey 2012:* www.turnerandtownsend.com/construction-cost-2012/_16803.html

Turner & Townsend (2013) *International Construction Cost Survey 2013:* www.turnerandtownsend.com/ICC-2013.html

Walsh, K. and Sawhney, A. (2004) *International Comparison of Cost for the Construction Sector: An Implementation Framework for the Basket of Construction Components Approach.* Report submitted to the African Development Bank and the World Bank Group, June.

Editorial comment

London is the most expensive city in the world to live in, beating New York, Paris and Sydney, according to a new study.

(Gander 2014)

After currency swings pushed Zurich to the top of the ranking last year, Tokyo has resumed its place as the world's most expensive city.

(EIU 2013)

We regularly see and hear news headlines that make claims about how expensive certain cities are compared to others, with 'league tables' published showing the most expensive, the cheapest and so on. Lead lines from two recent articles appear above. In fact the two studies show some very different results. The first, based on data from Expatistan.com, shows the five 'most expensive' cities to be London, Oslo, Geneva, Zurich and New York, while the second lists Tokyo, Osaka, Sydney, Oslo and Melbourne, with London not even appearing in the top 10. Even more puzzling is that, according to Expatistan.com (cited in Gander 2014), Caracas is the least expensive city of the 200 cities included in their research, while the Economist Intelligence Unit (EIU) lists the same city as the ninth most expensive out of more than 160 cities surveyed.

Other studies show similar differences. So why is it that the results can be so different and, in particular, why would a city such as Caracas show such wildly different results? The answer must lie in the methodology, and the opening words, "After currency swings", in the quote from the EIU, gives us at least part of the answer. If exchange rates are used to compare prices then differences in price levels are ignored and exchange rate fluctuations will change outcomes as currencies strengthen and weaken against others in open trading. Other aspects of the various methods will affect results to varying degrees; for instance, if the research is based on prices for a basket of goods and/or services, then both the content of the basket and the weighting of the individual items will be significant. For example, unleaded petrol or gasoline is a typical commodity that is bought and used all around the world; in Caracas, in oil-rich Venezuela, a litre of this fuel costs as little as two US cents, while in Oslo it costs around USD2.60. Regardless of how the prices are converted to USD, the real price difference will be huge. Other factors to be considered, then, are the importance (weighting) of petrol in the different studies and thus the impact that the price of petrol has on the final result.

The International Comparison Program (ICP) is an extensive exercise carried out by the World Bank that produces purchasing power parities (PPPs) that can be used to more realistically rank the wealth of countries and the standard of living of their people based on what they can buy with their money. It covers a wide range of expenditure across around 200 countries and enables the bank to identify the 'richest' and the 'poorest' countries based on measures such as real GDP per capita. It is not an exact measure of wealth or anything else, but it does allow countries to be categorized from the poorest to the wealthiest, and thus for the World Bank to target its assistance on the people most in need of it.

The ICP covers a wide range of goods and services and expenditure on construction is only one part of that. The methods used to collect and process data in the ICP are constantly evolving, and each pricing round has seen not only an increase in the number of countries covered but also substantial developments in the methods used. At the time of writing, the results of the 2011 round have just been published, and final details of the methods used have not been released. The following chapter, however, provides an overview of the processes used to gather construction cost data for the 2011 and earlier rounds and outlines some of the conceptual thinking behind the production of construction-specific PPPs.

4 The International Comparison Program and purchasing power parities for construction

Rick Best and Jim Meikle

Introduction

It is not uncommon to hear or read statements that suggest that people in some countries are living on just a couple dollars a day. "How can they survive?" we ask ourselves. The answer is that price levels are different between countries, and that is very much the case between wealthy industrialized nations and poorer developing countries. If one earns five local dollars a day and it costs only two to live, then survival at least is possible. The inconsistency lies in the use of money market exchange rates to convert amounts of money to the same currency and thus compare incomes and the price of goods in one currency with those in another. Comparing national economies via measures such as gross domestic product (GDP) using exchange rates to bring amounts expressed in various local currencies to a common base currency produces a view of relative wealth that can be very much distorted by price-level differences between countries.

International Comparison Program

The International Comparison Program (ICP) is intended to reduce these distortions by providing an alternative method for converting expenditure, costs and other amounts of money expressed in different currencies to a common base. The mechanism used is *purchasing power parity* (PPP). The ICP is an extensive data collection and analysis exercise that produces PPPs that are used for a variety of purposes, with one key purpose being to allow meaningful comparisons of national economies by reducing the distortions caused by money market or 'nominal' exchange rates and differing price levels.

In practice, there are different PPPs for different economic activities so that the price level of, for example, construction can be very different from, say, PPPs for machinery and equipment. In a later chapter, Meikle and Gruneberg present PPPs for a number of countries for the whole economy (GDP PPPs), construction, machinery and equipment, and different types of consumption. The general rules are that, as countries develop economically, their commercial exchange rates and PPPs tend to converge and that, in developing countries, internationally traded sectors – like machinery and equipment – have PPPs close to exchange rates, but largely domestic sectors – like construction – can have PPPs significantly different from exchange rates.

Origins of the ICP

The ICP began in 1968 as a joint effort between the United Nations Statistical Office and the University of Pennsylvania (Ahmed n.d.). The first round (1970) included only 10 countries; this had risen to 34 countries by the third (1975) round. Also during the 1970s, Eurostat began collecting data for similar purposes and in the early 1980s combined with the OECD to compare the GDPs of the various EU and OECD countries (OECD-Eurostat 2012a). The Eurostat and OECD exercises remain separate, but their results are used as part of the larger ICP. In 2000, the World Bank assumed responsibility for managing the ICP, and the 2005 round of the ICP covered 146 countries and economies. The 2011 round includes data from nearly 200 countries (World Bank 2013).

Components of the ICP

The ICP collects price data for a broad range of goods and services across a number of sectors of national economies. Construction is a key component, as it contributes significantly to the GDP of many countries and usually represents a major share of Gross Fixed Capital Formation (GFCF). The 2005 ICP round showed that for around two thirds of the 146 countries surveyed, construction expenditure (i.e. gross construction output) contributed between 9% and 18% of GDP, and in some it contributed more than 30% (McCarthy 2013). For example, in Australia in 2007–8, 'all construction' accounted for 16% of GDP and 57% of GFCF (ABS 2008). Other sectors surveyed include consumption (generally, the largest single component of GDP) machinery and equipment (the remainder of GFCF), government services, health and education.

Responsibility and organization

Today the ICP operates across six so-called regions, of which the OECD-Eurostat group is one; the others are the Asian Development Bank (ADB), the African Development Bank (AfDB), the Commonwealth of Independent States (CIS – a loose association of former Soviet republics), the United Nations Economic and Social Commission for Western Asia (UNESCWA) and the United Nations Economic Commission for Latin America and the Caribbean (UNECLAC). These agencies (e.g. the AfDB) oversee data collection and validation for their countries through national statistical offices, while the Global Office of the ICP sits within the World Bank. The Global Office of the World Bank oversees the whole operation and is responsible for producing global PPPs; it is advised by a Technical Advisory Group (TAG), an international group of experts in statistics, national accounts and economics, to manage the development and operation of the ICP. External consultants are contracted to advise on specific areas, including construction. The ICP infrastructure is large and complex but necessary for an exercise of this size and importance.

Methods for producing construction PPPs

In the context of the ICP, construction-specific purchasing power parities (CPPPs) are just one component of the PPPs produced to compare economies at the level of gross domestic product (GDP). Sector-level PPPs such as those for construction are specific to certain industries or parts of the economy and, in the case of construction, include PPPs at the level of three so-called basic headings: residential, nonresidential and civil engineering construction. Country-level PPPs represent a series of averages and, as such, should be used with caution when industry-level comparisons are the focus. In some cases, there is little difference between the country- or GDP-level PPP and the CPPP.

In the 2008 Benchmark PPP Results (OECD-Eurostat, 2008), the GDP PPP for Australia was 1.48 and the CPPP almost the same at 1.46 (base: US = 1.00); in contrast, the two PPPs for Greece were 0.701 and 0.492 and for South Korea, 785 and 539. In the latter cases, use of country PPPs rather than CPPPs to compare construction costs would produce substantially different results. Further differences are apparent if CPPPs for each of the basic headings are considered. Table 4.1 shows CPPPs for residential, nonresidential and civil engineering construction as well as for all construction for three EU countries with the average for 27 EU countries as a base of 1.00.

While the PPPs for 'all construction' do not vary greatly from the European average, there are substantial differences between PPPs for basic headings in some instances, for example between residential and engineering construction in the UK.

The basic process for the calculation of PPPs is illustrated by the well-known 'Big Mac Index' that has been published regularly for many

Table 4.1 Construction-specific PPPs for 2011 for selected EU countries (Eurostat 2012)

	Residential	Nonresidential	Engineering	All construction
Greece	0.850	1.042	0.824	0.907
Italy	0.846	0.835	0.952	0.864
UK	0.716	0.983	1.252	0.914
EU27	1.000	1.000	1.000	1.000

years (*The Economist* 2012). That index is produced by comparing the cost, in local currencies, of a standard hamburger in different countries. The ratio of the cost in one place with that in another gives a simple 'exchange rate' based on what can be bought with amounts of different currencies. For example, if a Big Mac hamburger in Australia costs AUD4.80 and USD4.20 in the US, then this implies an exchange rate of 1AUD = 0.875USD, as around 88 US cents buys as much hamburger (i.e. the same *volume* of output of product) as one Australian dollar. If the two hamburgers are identical, then it is assumed that they have the same value however it is measured, for example kilojoules, well-being or satisfaction.

A similar logic underlies all PPPs, whether the output or product that is being priced is a haircut or a tractor – or an amount of construction work. A hamburger implicitly embodies a variety of inputs that includes not only physical inputs such as bread and meat but also the labour required to make the product as well as other inputs such as rent. The selling price also includes profit. Construction work similarly embodies physical inputs, chiefly construction materials such as concrete and steel, but it also includes labour, equipment and overheads as well as other costs related to the realization of a building project, such as design fees.

The Big Mac Index was originally intended as a lighthearted attempt to demonstrate how currencies were under- or overvalued when compared to other currencies. One advantage it has is that the prices used are true purchaser prices; the cost of a Big Mac is reasonably standard within any given country, and the specification is fairly similar both within and between countries. The index is often criticized, however, because of the relatively small set of inputs that make up the product and the assumption that this set of inputs is representative of consumption generally in all locations. By contrast, the ICP surveys many thousands of products that it hopes will adequately represent all expenditure. In practice, this is more difficult than it may appear, as the products are still only a sample of what is purchased in any country, and there is no certainty that all of the items in the survey are actually representative of what is bought or consumed in any given location, and even where similar products are bought, they often vary greatly in detail and quality.

Construction presents an arguably greater level of difficulty. Two major concerns are the problem of collecting purchaser prices, the amount actually paid by clients, for construction projects and the problem of finding output that is similar enough to be comparable across locations and yet is reasonably representative of what is routinely built in the different countries.

In practical terms, acquiring true purchaser prices for construction projects is all but impossible, as final account (i.e. out-turn price) information is seldom available. Variations due to delays, design changes and a range of other factors generally mean that the final amount paid is not the same as the original contract sum, and even if such data is available, it will probably never be for projects that are essentially the same in many locations.

Furthermore, local practice in regard to allocation of risk and varying contract conditions can impact tender price versus out-turn cost, as the more scope there is for adjusting the final price, the lower the tender price. This is all part of the comparability/representativeness problem discussed in the preceding chapter.

In general terms, construction costs/prices can be measured in three ways: by pricing identical projects in each location, by pricing typical projects that can be modified to reflect some local variations or by pricing a representative basket of common inputs such as materials and labour or work items. The first option compares like with like, but it is very difficult to find projects, particularly buildings, that are designed and built in the same way in many places, so any project, whether real or hypothetical, chosen for such an exercise is unlikely to be representative of usual practice in every location.

The second option, through its flexibility in design and specification, allows for more representative projects to be priced, but this is only possible with some loss of comparability. In effect, what are priced are functionally similar projects that may vary considerably in their composition and technology. The third option is not project based; instead, a collection of typical inputs or resources is used to represent completed construction. The basket of inputs is priced and adjusted to represent purchaser prices by the addition of design fees, profit and overheads.

A number of methods have been suggested, and several have been extensively developed in order to arrive at practical ways of producing CPPPs while addressing the problems described. The main ones (which are described and discussed next) are:

- standard projects
- baskets of goods or resources
- baskets of construction components
- superficial areas.

It must be acknowledged that no method eliminates all the problems. Even if it were feasible to have identical projects constructed simultaneously in every country with accurate out-turn costs recorded, the problem of finding projects that would be representative of typical construction in every location would remain. Inevitably, any selected method will be a compromise between theory and practice. The problem is exacerbated because there is no 'correct' answer against which results can be compared. There are also practical limits in regard to how much effort participating countries can put into the whole exercise; less wealthy, developing countries simply cannot afford to commit scarce resources to extensive data collection exercises, and data quality suffers as a result. The ICP currently employs three methods to collect construction cost data, with the OECD-Eurostat division using a detailed standard projects approach, the CIS countries using what they call the Technical Resource Model approach and the remainder now using a basket of resources method that has replaced the previous basket of construction components (BOCC) method.

There is also the question of how representative projects are of all work in a sector or a country. Most projects used in this type of exercise are 'new' rather than 'work to existing', and price levels may vary across these different types of work.

The reasons for the use of different approaches are largely historical. The OECD and Eurostat groups have always used the bills-of-quantities approach and are reluctant to change their method. The CIS states similarly have a framework in place from Soviet times with reporting lines and databases already established, and they too prefer to continue with the method that they have in place.

Standard projects

The OECD-Eurostat PPP Program uses a set of hypothetical projects, including two civil engineering projects that are intended to represent 'typical' construction projects. These are described in detail by means of bills of quantities that are made up of hundreds of individual items of work that would be required to complete each project. These items include such things as '20 cm sand-lime brickwork + 11 cm facing brickwork, inclusive of pointing and acid cleaning'. Each item includes a quantity, and pricing the items means inserting a typical unit rate against each item and calculating an item price. Fig. 4.1 is an extract from a typical bill of quantities. The unit rates should include profit and both job-specific and general company overheads but not design fees or nondeductible tax such as VAT or GST. These costs do, however, form part of the overall price paid by clients for completed projects, and they are added to the sum of the item prices to reach an indicative total purchaser price for each project.

There are a number of problems, theoretical and practical, associated with the OECD-Eurostat method:

- Relatively few countries in the group routinely use bills of quantities for pricing construction work, and construction cost experts generally have to be hired to do the pricing.
- Those that do have different ideas of what should be included in or excluded from items.
- The number of items to be priced is quite large in spite of a significant reduction in the number of projects and work items in recent years; this contributes to implementation costs and generally means that only one set of prices is collected in each country.
- There are significant variations in 'typical' projects between countries within the group – in the case of houses, there are six regional variations that are priced by different countries (European, Nordic, Portuguese, North American, Australasian and Japanese single-family houses).

Item Specification	Unit	Quantity	Unit Price (national currency)	Total Price (national currency)
3. Masonry				
3.1 Ground floor double-skin external wall:	m²	257	17	4,369
• 20 cm sand-lime brickwork + 11 cm facing brickwork, inclusive of pointing and acid cleaning	m²	257	8	2,056
• Plastering				
3.2 Upper Floors double-skin external wall:	m²	413	18	7,434
• 11 cm sand lime brickwork +10 cm facing brickwork, inclusive of pointing and acid cleaning	m²	413	8	3,304
• Plastering				
3.3 Gable ends, 11 cm facing bricks, inclusive of pointing and acid cleaning	m²	625	18	11,250
3.4 Fair-finish 7 cm plaster block work	m²	585	7	4,095
Total				32,508

Figure 4.1 An extract from a typical bill of quantities (OECD-Eurostat 2012b)

- The projects may not be particularly representative of work carried out in each country.
- The method applies to new construction and does not directly address work to existing buildings and structures, which is a significant part of total construction expenditure in many countries.

Positive attributes include:

- Productivity differences between countries should be accounted for, as unit rates include the labour and equipment necessary to complete each priced item of work.
- The quantities attached to each work item reflect actual quantities in real projects, and thus the items are automatically weighted to reflect the expenditure shares of the various items; however, the weights only apply to these projects, which may or may not be representative.

Basket of components (intermediate prices)

In the 2005/06 pricing round for the ICP, construction cost data was collected using a specially developed basket of construction components (BOCC). There are two major challenges in adopting this approach: identifying suitable items to include in the basket and determining weights for the various items.

In the BOCC exercise, the first challenge was addressed through some fieldwork in a number of countries that involved inspecting construction work in progress and identifying items that were similar in each location. This was complicated by the need to identify items that were representative and comparable across the three basic headings or market sectors (residential, nonresidential and civil engineering construction). The BOCC that was finally assembled was a mixture of composite items (e.g. reinforced concrete columns, which included prices for concrete supply and placement, provision and installation of steel reinforcement and the materials and labour required for the construction of the necessary formwork or shuttering) and supply-only items (e.g. a quantity of cement). In addition, there were some labour-only items (see McCarthy 2013 for full details of the BOCC).

The BOCC method required weights at three levels (see following chapter for a more detailed discussion of weights within the CPPP process):

- W1: to aggregate the three basic headings (residential, nonresidential and engineering construction) to the 'all construction' level that would feed into GDP level PPPs
- W2: to aggregate the systems (substructure, superstructure, exterior shell and so on) to basic headings
- W3: these weights reflected the proportion of basic inputs (materials, labour and plant) into the various construction components in the BOCC

The W2 weights were to be obtained from in-country experts and determined by analysis of bills of quantities (BQs) for completed projects. In practice, this proved very challenging for the data collectors in many countries. While some countries provided reasonable data for W2 weights, many were not able to address this. There was also some confusion about W3 weights, with at least one region assuming that they had to produce these as well.

The implementation of the BOCC uncovered a number of problems both in the design of the basket and in the data collection phase. The inclusion of simple items, such as a cubic metre of ready-mixed concrete, and composite items, such as a reinforced concrete footing, meant there was potential for some double counting. In addition, the inclusion of a material (concrete), its components (sand, aggregate and cement) and a number of composite items that included concrete as a component made it impossible to relate results to national accounts categories. It also posed problems for the allocation of weights. Pricing required experts such as estimators and/or quantity surveyors, and the derivation for weights at the W2 level required historical data in the form of priced BQs, a resource that is rare, particularly in countries where the use of BQs is not common.

Observations:

- As implemented, the BOCC method was a mix of simple and complex items that lacked consistency with a mixture of input costs (e.g. quantities of sand and cement) and what were essentially purchaser prices for complete components (e.g. concrete columns and pad or spread footings).
- Implementation proved to be very difficult for those in the field, with particular problems encountered in collecting W2 weights.

The BOCC data was aggregated into PPPs using a country-product-dummy (CPD) method – methods for PPP calculation are beyond the scope of this book – refer to the ICP/World Bank website for more details (ICP 2014).

At face value, the component approach appears an attractive option, not least because it theoretically accounts for differences in labour productivity and capital/labour mix. In practice, however, it has not always been successful, at least in the international context.

The resource technological models (RTM) approach

The Commonwealth of Independent States (CIS) uses a variant of the basket-of-goods approach. Input costs are collected for construction labour and around 100 materials and energy items (e.g. diesel fuel), and these are fed into a number of cost models for a set of residential, nonresidential and civil engineering projects. Purchasing power parities are derived from these project costs. The model is represented generically as a linear equation that combines the cost of the various inputs as well as profit, taxes and other costs that comprise purchaser prices for a large number of project models (around 100 in total) that cover all three basic headings (ICP, 2005). The materials priced cover a large proportion (around 85%) of the materials typically used in construction in the CIS. Different sets of projects/models are identified that represent typical construction under the three basic headings in each country. PPPs for each basic heading are then computed as the price relatives of the unweighted geometric means of the total costs for each set of projects included for each basic heading in each country.

The equation for the RTM is

$$P_{mk} = \left[\left(1 + \frac{a}{100}\right) \sum_{j=1}^{m} PM_j QM_j + \sum_{k=1}^{l} PE_k QE_k + W \times F_k + \left(1 + \frac{S+A+B+P}{100}\right) \right]$$
$$\times I \times VAT$$

where:

P_{mk} is the predicted price that a purchaser would pay for a project of type k that includes m types of materials from the list of materials that have been specifically priced and a is the estimated proportion of other materials that are not on the list of materials that are specifically priced

PM and QM are the prices and quantities of the m types of materials from the list of materials that have been specifically priced

PE and QE are the prices and quantities of the l types of energy that have been specifically priced

F_k is the estimated number of work days required to build a project of type k

W is the daily wage rate

S is the percentage of wage costs paid by employers for their employees' social security

A is the rate of consumption of fixed capital (basically depreciation) expressed as a percentage of wage costs

B is other costs expressed as a percentage of wage costs

P is operating surplus expressed as a percentage of wage costs

I is the cost of 'engineering services' – this includes design fees, supervision, technical advice and similar costs expressed as a percentage of all other costs

VAT is the rate of value-added tax

This method is used only by the nine CIS countries (i.e. countries that were formerly part of the Soviet Union), and it is still in use, as its mechanisms for data collection and aggregation were well established when the USSR ceased to exist. The ICP Operational Manual (ICP, 2005) identifies a number of significant features of the method and notes a couple shortcomings:

- Countries are only required to collect supply costs/prices for materials, energy and labour – the rest of the information is estimated by industry experts, and this makes the cost of implementation relatively low. Some of the estimated variables (e.g. depreciation) are adjusted in consultation with in-country experts to reflect local conditions.
- As construction plant and equipment is usually owned by firms in the CIS countries, hire of plant is not included, and depreciation (variable A) represents consumption of fixed capital.
- Participants are asked to identify which of the 100 model projects are representative of typical construction in their country and whether projects of each type were constructed in the survey year. This is intended to produce sets of projects that are representative of typical construction in each country in each year.
- While the aim is to produce purchaser prices, much depends on the accuracy of the experts' estimates of variables such as labour/materials/energy splits, depreciation rates and profit margins.
- It is unusual to compute operating surplus or consumption of fixed capital as a percentage of wages cost rather than on basic project price.
- It is suggested that construction projects and methods were very similar across the former Soviet countries. However, with the breakup of the Soviet Union, that similarity may well decline over time and, given

the lack of homogeneity in projects and methods outside the CIS, there appears to be little potential for the application of such a method outside the CIS.

Other questions could be asked:

- It is not clear where project-specific overheads (Preliminaries) or general overheads are included – presumably in 'other costs' or perhaps as part of operating surplus. Both are significant components of the total project cost that a purchaser pays for a construction project.
- The daily wage rate must be an average of different types of labour weighted to reflect the varying amounts of each level of labour employed across a national industry; this may have been a simpler task in centrally controlled economies than it is in market economies.
- Estimates of construction time will have an influence on predicted project costs, yet they can be nothing more than estimates of a target that is regularly missed in practice, with many projects taking considerably longer to complete than predicted.

ICP construction PPPs – the new approach for 2011

In 2009, a review of the BOCC method was initiated by the ICP Global Office in response to the difficulties that had been encountered in the implementation of the BOCC in the 2005 ICP round. From that review, a revised method based on a basket of resources or inputs (labour, materials and so on) was developed and a new survey devised to collect the cost data for the 2011 round.

The 2011 approach differs from the BOCC in that it collects contractors' costs (i.e. what they pay to purchase) for a range of materials, manufactured components and products, including energy products (petrol, diesel and electricity), as well as hire rates for items of construction plant and various categories of labour (for full details see ICP, 2011). While the BOCC did include some basic input items such as sand and cement, it was mostly focused on composite components such as reinforced concrete columns and areas of a particular finish such as cement render that are assembled or constructed on site. One of the criticisms of the BOCC was the mix of basic and composite items and the perceived double counting of items such as cement that appeared both as a basic input and as a component of items such as reinforced concrete and cement render.

In the 2011 approach, the prices paid by contractors to their suppliers are collected, via a survey, for 38 materials and products (including some manufactured components such as wash basins), the hire of five types of plant (which may be ownership costs incurred by contractors who own their own plant) and seven categories of labour. These prices are both annual average prices and national average prices for the survey year. The survey is typically completed by construction experts and, in most cases, only a single response is obtained from each country, although some countries do provide multiple responses. While this approach is not used in either the CIS or OECD/Eurostat countries in the 2011 round, some OECD/Eurostat countries completed the survey to assist with linking their data into the general ICP exercise.

In the 2011 round, data was collected for contractors' markups (profit and overheads) and professional fees with the intent that appropriate percentages would be added to the various resource prices in order to bring contractors' costs to purchaser prices (i.e. what the construction client would pay in total for the construction work that they purchase). In practice, no adjustment was made to prices for these factors due to the poor quality of the data collected (ICP, 2013).

Survey respondents were also asked to indicate the 'importance' of each material item in the basket for their country, that is which of the three basic headings (residential, nonresidential, civil engineering) was each item relevant to and was thus 'important' (i.e. readily available and commonly used in that type of construction). In some countries, some materials may well not be 'important' in any of the basic headings. 'Importance' equates generally to 'representativeness', a term used commonly in the ICP in the past but now largely replaced by 'importance'. Items that were deemed 'not important' under a basic heading in a given country might still be priced but their prices omitted from basic headings in which they were deemed 'not important'.

The survey data is then used to produce price relatives for each priced item compared to a base (typically the US price), which are then aggregated to produce PPPs for each resource type (i.e. materials, labour and plant). Each resource type (i.e. labour, materials and plant) is then weighted according to the proportion of construction cost that each type represents within a given basic heading. Resource PPPs are calculated for materials, labour and plant under the three basic headings, and these, in turn, are aggregated to produce PPPs for those basic headings. Finally, the three basic-heading PPPs are aggregated to produce construction PPPs, with each of the basic-heading PPPs weighted according to the proportion of total construction that each represents in each country. A step-by-step outline of how CPPPs might be produced from input price data is presented in the following chapter.

The use of indicators of 'importance' to select material items permits the identification of a basket for each basic heading that reflects common usage and does not necessitate prices for 'special' items that are not commonly available. This means, of course, that baskets are not identical across countries or basic headings, unlike the BLOC method described in the previous chapter. There are other differences.

The use of unweighted (or equally weighted) price relatives to produce resource PPPs is a fundamental difference between the ICP method and the BLOC method. Other differences between the two methods are the use of projects and specific locations in the BLOC rather than construction aggregates (basic headings and all construction) and national averages in the ICP. The BLOC method is one that allows practitioners to compare projects across specific locations; the ICP method is a statistical device to allow the production of national sectoral/work type PPPs.

Next steps

The ICP and PPPs are less well known and understood than they should be, particularly, for example, in the construction sector. If they are to be more widely accepted and used, three main issues need to be addressed: the time to produce results needs to be reduced; the frequency of results needs to be increased; and subnational PPPs at a major city or regional level are required.

Results from the ICP take a long time to emerge. The results of the 2005 survey were published in 2008, and the results of the 2011 survey started to appear in 2014. There are good reasons for this kind of delay in an exercise covering 200 countries of very different characteristics and capabilities. The surveys require data collected specifically for the exercise; the logistical issues involved are enormous; survey responses are delayed; data needs to be checked, validated and aggregated; there are implications for the relative sizes and rankings of economies; results need to be accepted by all concerned; and ICP PPPs need to be reconciled with the PPPs that have been used in official data sets while the ICP data is being produced. But somehow the time to publication must be reduced. This will almost certainly involve compromises and future revisions, and such actions will require international agreement, which will itself take time.

Just as three years for results to emerge is too long, every five or six years for PPP rounds is also too long. The OECD and Eurostat undertake PPP surveys more frequently (annually, in the case of Eurostat), and OECD and Eurostat data is published more promptly than the ICP, but these are relatively well-funded exercises and involve fewer countries and, for construction, involve the use of national experts. Eurostat is also conscious of the regional dimension, particularly in its larger members, but has still to address it. Industry observers, of course, would like specific locations, capital and main cities. Until these issues are addressed, adoption and use of PPPs will probably continue to be limited.

In conclusion, if PPPs – and particularly construction PPPs – are to be more widely used, they need to be more accessible, but the limits on their use and usefulness need to be set out clearly. Construction PPPs are produced to adjust broad categories of output in various currencies to common units of value, and they typically represent annual average and national average price levels. They should not be used to adjust particular projects or construction work in particular locations.

References and Further Reading

ABS (2008) *Australian System of National Accounts, 2007–08*. Australian Bureau of Statistics: www.abs.gov.au/AUSSTATS/abs@.nsf/DetailsPage/5204.02007–08? OpenDocument

Ahmed, S. (n.d.) *History of the International Comparison Program*. Available at: http://go.worldbank.org/NSHJORNI10: Accessed May 2013.

BCIS (2009) *2009 BCIS International Cost Elements Report* (UK: Building Cost Information Service).

The Economist (2012) *Big Mac Index*: www.economist.com/node/21542808: Accessed May 2013.

EIU (2013) *Worldwide Cost of Living 2013*. The Economist Intelligence Unit. www.iberglobal.com/files/cities_cost_living_eiu.pdf

Eurostat (2012) *Purchasing Power Parities (PPPs), Price Level Indices and Real Expenditures for Esa95 Aggregates*: http://ec.europa.eu/eurostat/en/web/products-datasets/-/PRC_PPP_IND: Accessed June 2013.

Gander, K. (2014) London comes top in study of the world's most expensive cities to live in. *The Independent*, 6 January. www.independent.co.uk/news/uk/home-news/london-comes-top-in-study-of-the-worlds-most-expensive-cities-to-live-in-9042233.html. Accessed 28 January 2014.

ICP (2005) *ICP Operation Manual*. International Comparison Program: http://siteresources.worldbank.org/ICPINT/Resources/ICPOperationalManual2005.pdf

ICP (2011) *A New Approach to International Construction Price Comparison*. International Comparison Program: http://siteresources.worldbank.org/ICPINT/Resources/270056–1255977254560/6483625–1273849421891/110622_ICP-OM_Construction.pdf

ICP (2013) *Technical Advisory Group: Minutes of the Eighth Meeting, May 20–21, 2013*. Available at: http://siteresources.worldbank.org/ICPINT/Resources/270056–1255977007108/6483550–1257349667891/6544465–1366815793411/130701_ICP-TAG08_Minutes.pdf

ICP (2014) *ICP 2011*. International Comparison Program, World Bank: http://siteresources.worldbank.org/ICPEXT/Resources/ICP_2011.html

McCarthy, P. (2013) Construction. In: World Bank. *Measuring the Real Size of the World Economy: The Framework, Methodology, and Results of the International Comparison Program – ICP* (Washington, DC: World Bank).

OECD-Eurostat (2008). *2008 Benchmark results* (OECD Publishing): http://stats.oecd.org/Index.aspx?DataSetCode=PPP2008

OECD-Eurostat (2012a) Purpose of the Eurostat-OECD PPP Programme. In *Eurostat-OECD Methodological Manual on Purchasing Power Parities* (OECD Publishing): www.oecd.org/std/ppp/manual

OECD-Eurostat (2012b) Construction. In: *Eurostat-OECD Methodological Manual on Purchasing Power Parities* (OECD Publishing): www.oecd.org/std/prices-ppp/12-3012041ec014.pdf

Walsh, K. and Sawhney, A. (2004) *International Comparison of Cost for the Construction Sector: An Implementation Framework for the Basket of Construction Components Approach*. Report submitted to the African Development Bank and the World Bank Group, June.

World Bank (2013) *Measuring the Real Size of the World Economy: The Framework, Methodology, and Results of the International Comparison Program – ICP* (Washington DC: World Bank).

Chapter 4 Appendix

Introduction

In this section, the process of creating a basket-of-goods based pricing instrument is discussed. The process described is based on that followed by the authors during the development of a preliminary version of a new data collection instrument for the ICP in 2010. The key points covered are the selection of the items included in the basket and the wording of specifications/item descriptions to facilitate the collection of comparable prices from different locations.

General characteristics

Once a decision had been made that a basket of inputs or resources would be used to collect construction cost/price data for the ICP, the authors had to consider what items would be included in order that the most useful data set could be assembled. Basic inputs to construction are materials, labour and plant/equipment, but the overall cost to construct a project includes other factors such as design and other professional fees, project-specific overheads (referred to as Preliminaries in the UK and Australia, for example), general or company overheads and contractors' profit. As the ultimate aim of the basket-of-inputs approach was to collect the equivalent of prices paid by clients for construction, additional information apart from the cost of the basic inputs was needed. However, the initial task was to select a basket of items of materials, labour and plant that would be representative of a construction inputs generally. The authors considered that the key components should comprise:

- basic materials such as sand and aggregates
- manufactured goods such as ceramics and paint that are incorporated into buildings in a similar fashion to basic materials
- complex items such as windows and electrical components that are installed as units
- labour required to fabricate and fix materials on site
- supervisory labour
- hire of plant and equipment required in the construction process.

To be representative and comparable, the items selected for the basket of resources or inputs needed to be:

- in common use in many locations
- able to be specified in clear terms so that very similar items or equivalent will be priced in all locations
- typically used in quantities that make them significant factors in total construction cost of projects.

The aim is that the prices collected will be for commonly available items; in no cases should prices for specially ordered items be provided. Where there are small differences in dimensions, adjustments can be made: for example, if the 'standard' brick thickness in one country is 100mm and, in another, 110mm, prices can be adjusted by 10%. Where there are differences in materials, the notion of 'function' should be invoked: if clay common bricks are standard in one country and sand cement common bricks are standard in another, they can be taken as equivalent to each other.

The authors were aware that an important aspect of the exercise was to keep the cost of data collection to a minimum, and thus the basket, while needing to be comprehensive enough to represent the industry, had to be such that in-country pricing could be done reasonably quickly and cheaply. Generally speaking, the more cheaply data collection can be done, the more opportunity there is to obtain more than one set of price observations in each country. It is also important that as many product groups as possible are represented so that the spread of items includes as many as possible of the main groups that feed into construction; this allows the dataset to be used for other purposes within national statistical offices.

Review of previous studies

The authors looked at the items included in a number of baskets used in both one-off studies and some regular pricing exercises carried out by consulting firms. Commonly occurring items were identified, and these formed the basis of the new basket. Other items were included based on personal experience with previous studies and identified gaps in SIC (Standard Industrial Classification) categories (see ICP, 2011a for an explanation of the relevance of this aspect of the exercise). The initial results of the analysis are shown in Table 4A.1. Note that this table shows only materials and products; labour items are dealt with separately.

From the initial analysis, a summary was derived (see Table 4A.2). Some similar items were combined, for example various specifications for aggregate in concrete. The summary comprised 24 materials and products. Table 4A.3 shows the summary items arranged by SIC groups.

Table 4A.1 Analysis of material items in previous baskets. BLOC. Basket of Locally Obtained Commodities (Best, 2008); BoG: Basket of Goods (DLC, 2003); BOCC: Basket of Construction Components – current ICP method; ICEC: International Cost Engineering Council (ICEC, n.d.); G&T: Gardiner & Theobald International Construction Cost Survey (G&T, 2010); CIS: Resource Technological Models from the CIS (CIS, 2005); BICS: Building Cost Information Service (BICS, 2006); F+G: Faithful+Gould (2010)

MATERIALS AND PRODUCTS	BLOC	BoG	BOCC	ICEC	G&T	CIS	BCIS	F+G	Frequency
Softwood (carpentry)		x			x		x		3
Sawn lumber						x			1
Small aggregate		x							1
Coarse aggregate		x						x	2
Aggregate (general)			x	x	x	x	x		5
Sand		x	x		x	x		x	5
Emulsion paint	x	x	x				x		4
Oil paint		x							1
Paints						x			1
Windows								x	1
Vinyl floor tiles/sheet		x				x		x	3
Plastic cold water pipe		x	x						2
PVC pipe						x		x	2
Exterior grade plywood		x					x	x	3
Interior grade plywood (joinery)		x							1
Plywood for formwork (formply)	x		x						2
Chipboard flooring		x				x			2
Timber window		x							1
Joinery unit – wardrobe	x								1

(Continued)

Table 4A.1 (Continued)

MATERIALS AND PRODUCTS	BLOC	BoG	BOCC	ICEC	G&T	CIS	BCIS	F+G	Frequency
Single internal door exc hw, frame	x								1
Single fire door inc. hardware, frame	x	x						x	3
Timber door (internal)	x					x			2
Carpet	x								1
Cement	x	x	x		x	x	x	x	7
Ready-mixed concrete – various MPa (N/mm^2)	x	x	x		x	x	x	x	7
Sandstone	x								1
Clay roof tiles		x							1
Common bricks		x			x	x	x		4
Facing bricks		x							1
Ceramic wall tiles	x	x				x			3
Ceramic floor tiles	x					x			2
Clay floor tiles		x							1
Clay drain pipe		x				x			2
Plasterboard	x	x		x			x	x	5
Plaster					x		x		2
Hollow blocks	x	x	x	x	x	x	x	x	8
Precast paving		x							1
Precast concrete units	x		x					x	3
Solid blocks		x				x			2
Concrete roof tiles		x							1
Insulation						x	x		2
Aluminium-framed window	x		x			x			3

Float glass 6mm thick		x	x		x	x	x	x	6
Double-glazed unit		x							1
Toughened glass	x								1
Rigid air conditioning duct	x					x		x	3
High-tensile reinforcing bar	x	x	x		x		x	x	6
Plain reinforcing bar		x							1
Structural steel	x	x	x	x	x	x	x	x	8
Metal deck roofing						x		x	2
Cable tray				x					1
Cast-iron drain pipe	x	x				x		x	4
Steel pipe				x		x		x	3
Copper pipe 15mm dia.	x	x	x					x	4
PVC sheathed electrical cable	x	x	x	x		x			5
Flexible air conditioning duct	x								1
White hand basin		x				x			2
White WC suite	x	x				x		x	4
Vanity basin	x								1
Hot water boiler (93kw)								x	1
Electric wire, 4mm² cross section								x	1
20mm rigid steel conduit								x	1
Mercury vapour light fitting, 400 watt								x	1
Fluorescent light fitting, 4 lamp, 40 watt						x		x	2
Downlight								x	1
No. of items	27	35	14	8	11	28	15	25	

Table 4A.2 Summary of material based on frequency of inclusion in previous exercises

MATERIALS	FREQUENCY	NOTES
Sawn timber	4	Also part of formwork
Aggregate	8	Varying specifications
Sand	5	Generic item
Paint	6	Spec ranges from generic "paint" to particular types (e.g. 'emulsion')
Windows	6	Some are frames only, fully glazed, single and double glazed
Vinyl floor coverings	3	Tiles and sheet
Plastic pipe	4	Varying specifications
Plywood	6	Various types: internal, external, formply
Timber door	6	Various types, inc/exc frame and hardware
Cement	7	In some cases as a component of concrete not as a discrete item
Concrete	7	Varying strengths
Bricks	5	Mostly commons (4)
Ceramic tiles	5	Wall and floor – clay tiles not included
Plasterboard	5	Also plaster occurs twice
Concrete blocks	10	Mostly hollow (8)
Glass	7	Mostly 6mm clear float (6)
Reinforcing steel	7	Mostly high tensile (6)
Structural steel	8	
Roof coverings	3	Various types – no clear choice
Cast iron/steel pipe	7	4/3 split
Copper pipe	4	Small diameter
Electrical cable	5	
Basin	3	Hand basin/vanity basin
WC suite/pan	4	

Commentary on previous exercises

Each of the exercises listed in Table 4A.1 has its own characteristics, and it is worth commenting on the various approaches.

The BLOC (Basket of Locally Obtained Commodities)

The basket used in this exercise was derived from analysis of a priced bill of quantities for a completed hotel project located in suburban Sydney. Cost-significant items in each trade were identified, and a weighted basket of goods was derived. The basket was representative of a particular building type in Australia. It included 24 material items; during data validation, one (prestressing steel) was discarded, as the price data collected was very variable, and it appeared that few respondents were familiar with a supply price for a material that is typically included in all-in subcontractor prices for prestressing that include supply and fixing of all relevant materials and

Table 4A.3 Material items in summary arranged by SIC code

SIC CODE	INDUSTRY GROUP	SELECTED MATERIALS AND PRODUCTS
14	Other mining and quarrying	Aggregates and sand for concrete
20	Wood and wood products	Sawn timber, plywood, timber door, timber window (†)
24	Chemicals	Paint
25	Rubber and plastic products	Vinyl floor coverings, plastic pipe
26	Glass, ceramic, clay and cement products	Cement, ready-mix concrete, plasterboard, bricks, concrete blocks, float/sheet glass, window glass (†), ceramic tiles, basin, WC, roof coverings (†)
27/28	Metal products	Reinforcing steel, structural steel, cast-iron drain pipe, copper pipe, roof coverings (†),
31	Electrical machinery	PVC insulated cable

(†) means the product may be part of more than one code, as it is a composite product such as a timber-framed window with glass infill, or it depends on the item specified (e.g. roof coverings might be clay tiles or metal sheet)

accessories. Best (2008) contains a detailed discussion of how each of the items in the basket performed 'in the field' as well as details of how each item was selected and weighted. The basket included a number of items related to building services. It was not intended to be representative of any other type of construction (such as civil engineering construction or residential building)

The BoG (Basket of Goods)

This basket was produced by Davis Langdon Consultancy and tested across a number of European countries. Material selection was based on personal experience. It included 35 material items, of which 20% related to building services. It was intended for use in the European context, and the choice of materials does to some extent reflect that, for example concrete and clay roof tiles as the only roof-covering materials.

The BOCC (Basket of Construction Components)

This basket was used in the 2005 ICP pricing round. It contains both simple or basic materials as well as composite or complex work items. In this analysis, only the basic items were considered, although it should be noted that many of the composite items included basic materials. Selection of items was based on extensive research in a number of countries where the designers of the basket looked at various materials and components or assemblies of materials (e.g. concrete footings and columns). The basket contained only one item related to services, and that was difficult to price with any confidence.

The ICEC location factor model

This initiative by the International Cost Engineering Council was intended to collect a relatively small set of costs for a model project comprising a process plant and a small building. The survey form included composite and basic items and was weighted towards engineering construction of a certain type. It was included in this analysis, as it did contain some of the commonly occurring items such as concrete, gravel and sand. It should be noted that the ICEC asked for members to provide prices online, and almost no-one responded; hence, there is no information available regarding how well or how poorly the survey instrument worked.

Gardiner & Theobald

Gardiner & Theobald is an international firm of construction consultants. They publish the results of a regular cost survey of international construction. Their list of key materials is quite short, comprising only 11 items, of which none are related to services. Although the list is short, it contains key materials, and it is the choice of these items that is of interest here. It is not a basket as such but more a set of indicative prices that are published as much as a marketing exercise as for any other purpose.

CIS – the resource technological models

This method is based on a list of 66 items that includes basic and composite items plus three energy items. Only items that roughly correspond to the other baskets in the analysis were included, for example concrete and cement are included while foundations and external wall panels for residential buildings are not. The list is based on standard products that are a carryover from the centrally controlled economies of the Soviet era. As such, the statistical offices in the various member states are able to price the items in accordance with established common specifications. Once again, the significance here is in the key items included that reflect similar choices in other baskets. The items in the list are intended to represent construction typical of the three basic headings, with a number of items being specific to civil engineering construction (e.g. railway ties, rails and cable). Some 16 of the 63 material items are services related, although in some cases the exact nature of items is a little unclear due to very brief specifications and some translation anomalies.

BCIS – Building Cost Information Service – Asia building construction survey

This is another study that included input costs for 15 materials, but in this case no basket prices were compiled. Materials costs were not the primary focus of the study and were included to allow indicative comparisons of relative construction costs in a number of Asian countries. Once again, the significance is in the choice of items that the BCIS consider to the key items. No services items were included.

Faithful+Gould – international construction intelligence

This is a regular survey carried out in 32 countries. It includes a list of 25 material inputs. How the price data is used is not clear, as Faithful+Gould do not publish anything that obviously uses the data. They do publish a parity index based on a model building, and the survey collects a range of data, including some composite rates for elemental items, inflation, labour costs and details of procurement arrangements.

Summary

Several key factors emerge from the foregoing analysis.

- A number of materials appear in many of the lists, with some that are more or less ubiquitous, such as concrete/cement, structural and reinforcing steel and several others.
- Some of the lists are produced for purposes other than the population of a basket to be used for comparative purposes, and of these lists some are quite short, being intended as simple indicators of comparative costs between locations.
- Although building services represent between 20% and 50% of building costs, services-related items are poorly represented or even ignored completely.

Selecting materials for the basket

With the summary from Table 4A.2 as a starting point and drawing on the authors' personal experience with a number of earlier pricing exercises a preliminary list of materials for inclusion in the basket was developed. The first cut is shown in Table A4.4, with explanatory comments added by the authors. The 23 items listed in Table 4A.4 cover eight SIC groups; to maximize the utility of the data gathered, coverage of more SIC groups is useful. With that in mind, the following items were suggested for inclusion in addition to those shown in Table 4A.4:

- carpet (SIC 17 – Manufacture of textiles)
- petrol (gasoline) and diesel (SIC 23 — Petroleum products)
- pumps and fans (SIC 29 – Manufacture of machinery and equipment).

Inclusion of these items would increase the coverage from 8 SIC groups to 11.

Carpet, although appearing only once in the analysis summarized in Table 4A.1, is in common use, and in the study based on the BLOC (Best,

Table 4A.4 First cut of list of materials for inclusion in a basket of resources

SIC CODE	MATERIAL	UNIT	NOTES
14	Aggregate	tonne or m³	Suggested specification: "Coarse aggregate for use in concrete" – either unit can be used to suit local conventions – eventually data will need to be normalized for one unit – this can be done centrally or in-country
14	Sand	tonne or m³	"Sand for use in concrete and render" – ditto
20	Sawn timber	m³	Softwood for framing is the most likely specification. Price per m3 avoids problems with different standard sectional sizes and nominal versus actual sizes for dressed timber
20/26	Windows	each	It may be advisable to offer two alternatives: metal frame and timber frame – we can insert dummy prices where necessary
20	Plywood	m²	Thickness and intended use determine price
20	Timber door	each	Excluding frame and hardware. Specify by overall size (approx.) and location (internal use) – no fire resistance rating. Flush finish ready for paint
24	Paint	litre	'Emulsion' or 'water-based' – suitable for interior render/plaster surfaces
25	Vinyl sheet	m²	The choice is between tiles and sheet – in Australia, there is a substantial difference between the supply cost for sheet and tiles, with sheet much cheaper than tiles
25	Plastic pipe	m	Size and intended use determine price
26	Cement	tonne	Ordinary Portland cement for general building work
26	Concrete	m³ or tonne	One midrange strength is sufficient, say 25MPa (25N/mm2)
26	Bricks	m² or each	If priced per brick, local dimensions are required. Common bricks only – facings vary too much in quality and cost and aren't used much in some places
26	Ceramic tiles	m²	Pricing by unit area includes various tile sizes. Specify either wall or floor, as thickness varies considerably
26	Plasterboard	m²	Specify approximate thickness (there are local variations in standard thicknesses) and intended use. Also referred to as sheetrock or gypsum board
26	Concrete blocks	m² or each	Area rates would be preferable, as block sizes vary considerably from place to place. Hollow blocks seem most common

(Continued)

SIC CODE	MATERIAL	UNIT	NOTES
26	Glass	m²	Plain float glass – nominal size 6mm thick
26	Ceramic wash basin	each	Suggest inclusion of only one item of sanitaryware – more price variability with toilets than basins
27	Iron/steel pipe	m	Specify nominal diameter and intended use (e.g. waste-water plumbing)
28	Reinforcing steel	tonne	High-strength bars only
28	Structural steel	tonne	Important to give indication of section size and to include 'fabrication and delivery to site'
28	Metal roof sheeting	m²	This appears to be the most common generic roof covering – tiles, regardless of material, are not used everywhere
28	Copper pipe	m	Small diameter for water supply plumbing
31	Electrical cable	m	PVC sheathed and insulated, copper core. Specify 'for large quantities', as price varies a lot depending on volume

2008) was priced consistently with just a single outlier from a dataset of 50 observations across six locations. The fuel items appear only in the CIS list (and one, unleaded petrol, in the BCIS list) but were included, as it was considered that the data may be useful as a proxy for construction equipment. Electrical equipment constitutes a significant cost component in mechanical and hydraulic services; thus, the fan and pump items filled a gap in the basket as well as adding another SIC group to the mix.

Draft list of materials with specifications

Table 4A.5 shows a second draft of a list of items for pricing, including units and draft specification, with notes for each item. There are 29 items in the list, including two fuel items.

The intention, where there are options for pricing (e.g. aggregate may be priced by the tonne or per m³; bricks and blocks may be priced per unit/each or per m²), was that respondents could supply some additional data to allow others to convert prices to common units. A glossary of terms would give clear definitions of terms where it might be expected that there would be differences in regional terminology. Additional instructions were deemed necessary to cover pricing parameters such as the exclusion of value-added tax (VAT) or goods and services tax (GST), inclusion of delivery to site, pricing 'supply only' not 'supply and fix' and so on.

Table 4A.5 Proposed basket of resources (materials and fuel only)

SIC	ITEM	SPECIFICATION	UNIT	NOTES
14	Aggregate	Coarse aggregate (crushed rock, gravel) for use in concrete	tonne	If pricing in m^3, please give indication of density (kg/m^3) of typical aggregate in your location
14	Sand	Sand suitable for use in concrete and render	tonne	If pricing in m^3 please give indication of density (kg/m^3) of typical aggregate in your location
17	Carpet	80/20 wool blend carpet, average commercial quality	m^2	Typical for use in mid-range hotel or residential
20	Sawn timber	Softwood sections for use in general carpentry (wall framing, roof framing and similar)	m^3	Typical sections such as 100mm × 50mm
20/26	Timber window	Approx. 500mm × 500mm, timber frame, glazed with one fixed pane of 6mm plain glass	each	Average cost for domestic or commercial use – fixed pane hence no hardware included
20	Plywood	15–18mm-thick plywood for basic interior use	m^2	Do not price for water resistance, decorative finish or exterior grade
20	Timber door	Timber door, interior use, nominal overall size 2000 × 800 × 35mm, flush panel for paint	each	Excluding hardware, frame and all other accessories
23	Petrol/ gasoline	Standard-grade unleaded petrol/gasoline	litre	
23	Diesel fuel	Standard-grade diesel fuel	litre	
24	Paint	Low-sheen emulsion (water based) paint suitable for interior plaster and render	litre	Commercial quality and quantity
25	Vinyl sheet	2mm-thick vinyl sheet – average quality	m^2	Commercial quality and quantity
25	Plastic pipe	150mm-diameter plastic drain pipe	m	Straight lengths of plain pipe suitable for stormwater (rainwater) in ground
26	Cement	Ordinary Portland cement for general building work	tonne	Contractor quantity
26	Concrete	25MPa (25N/mm2) ready-mixed concrete delivered to site	m^3	Medium size project quantities
26	Bricks	Common bricks suitable for rendering or painting	m^2	If pricing in other units please specify dimensions of bricks

(Continued)

SIC	ITEM	SPECIFICATION	UNIT	NOTES
26	Ceramic tiles	300mm × 300mm (approx.) ceramic floor tiles – average quality	m^2	Commercial quality and quantity
26	Plasterboard	13mm-thick paper faced gypsum board	m^2	Also known as sheetrock – suitable quality for basic interior use (wall and ceiling linings)
26	Concrete blocks	Plain hollow blocks – basic quality	m^2	If pricing in other units please specify dimensions of blocks – no reinforcement or core filling
26	Glass	6mm-thick plain float glass	m^2	Suitable for basic glazing in frames
26	Wash basin	White-fired clay hand basin – wall or bench mounted	each	Commercial quality and quantity – no tapware or plumbing
26/28	Metal window	Approx. 500mm × 500mm, aluminium frame, glazed with one fixed pane of 6mm plain glass	each	Average cost for domestic or commercial use – fixed pane hence no hardware included
27	Iron/steel pipe	100mm-diameter steel or cast-iron drainage pipe	m	Average quality, interior use
28	Rebar	High strength steel bar reinforcement 12–24mm diameter	tonne	Medium size project quantities
28	Structural steel	Standard steel sections 150–300mm max cross sectional dimension	tonne	Including fabrication and delivery
28	Metal roofing	Simple profile galvanized steel roof sheeting	m^2	No colour treatment
28	Copper pipe	Small-diameter (15–20mm) pipe suitable for hot and cold water reticulation	m	Straight lengths of plain pipe in medium size project quantities
29	Electric pump	Electric pump, flow rate 10 litres/second, head pressure 150KPa	each	For pumping water, temperature range 5–80°C
29	Fan	Electric exhaust fan, flow rate 1000 litres/second, head pressure 250Pa	each	For interior installation
31	Electrical cable	PVC sheathed and PVC insulated cable suitable for domestic circuits	m	Medium size project quantities

Table 4A.6 Labour items from previous baskets

LABOUR	BLOC	BoG	BOCC	ICEC	G&T	CIS	BCIS	F+G	Frequency
Trade/ skilled Level 1	x								1
Trade/ skilled Level 2	x								1
Trade/skilled Level 3	x								1
Unskilled		x	x	x	x		x		5
Semiskilled	x	x			x		x		4
Skilled		x	x	x	x		x		5
Site supervisor/ engineer	Prelims†	x		x			x		4
Labour (average)						x			1
Structural steelworker								x	1
Bricklayer/ Mason								x	1
Carpenter								x	1
Roofer								x	1
Labourer								x	1
Sheetmetal worker								x	1
Plumber								x	1
Electrician								x	1
Designer/ draftsman				x					1
No. of items	5	4	2	4	3	1	4	8	

†Prelims: supervision included in Preliminaries percentage add-on

Labour items

Table 4A.6 shows the results of the authors' analysis of labour items included in previous baskets. Table 4A.7 shows the summary; once again some items are combined, e.g. 'Skilled', 'Tradesman Level 1' and 'Electrician' were considered to be the same in this context.

Methods vary from the CIS approach, which uses just a single average rate for labour, to the quite detailed breakdown used by Faithful+Gould.

Table 4A.8 shows the labour items proposed for inclusion in the basket.

Plant and equipment

While materials generally account for a large proportion of total construction costs, plant and equipment used on site can still be a significant cost in building work and are generally more significant in engineering construction. Plant and equipment are costed in different ways according to country and the manner in which items are sourced (i.e. owned by contractors or hired as needed). For example, in Australia, it is common practice for large

Table 4A.7 Frequency of labour items in previous baskets

LABOUR TYPE	FREQUENCY	NOTES
Skilled	10+	Breakup varies – definition varies by location
Semiskilled	6	Breakup varies – definition varies by location
Unskilled	6	
Supervisor	4	Includes one instance in Preliminaries

Table 4A.8 Proposed labour items for inclusion in the basket

LABOUR TYPE	UNIT	NOTES
Skilled	hour	Specialist tradesperson requiring expert knowledge, e.g. electrician, plumber
Semiskilled	hour	Trained worker utilizing skills acquired on the job, e.g. bricklayer, painter, plasterer
Unskilled	hour	General labourer performing basic tasks such as unloading and moving materials, cleaning up and similar
Supervisor	hour	Onsite supervision, e.g. foreman, leading hand, site manager

items of plant that are used by more than one trade (e.g. a tower crane that hoists all manner of materials off trucks and to storage areas or to working levels for installation) to be costed as part of a Preliminaries package (see what follows). Preliminaries also include other site-/job-specific costs that the contractor incurs in the course of building a project apart from direct costs (e.g. labour and materials). They include insurances, site amenities and site water and electricity supply. In the USA, however, craneage is usually treated as a separate cost centre. If a crane is hired, then daily/weekly/monthly costs are incurred; if the contractor owns the item, then a time-related cost of ownership is charged against the job which includes depreciation, maintenance, fuel, operator and so on.

The authors' analysis of previous baskets revealed that plant and equipment were seldom addressed. The BOCC included just two items, the ICEC model included crane hire and two included energy (fuels and electricity) in some way, while the rest did not include plant directly in any way. The BOCC also included plant costs in composite rates for components; however, there were no details available to verify how well that worked in practice.

While plant and equipment may be dealt with by other means, it was proposed that three typical items be included in the survey. At the time of the development of the new basket, the precise method for dealing with plant was still unclear, but it was thought prudent to test the viability of collecting direct prices for some items. Table 4A.9 shows the labour items proposed for inclusion in the basket.

Table 4A.9 Proposed plant items for inclusion in the basket

ITEM	UNIT	NOTES
Backhoe	day	Suitable for excavating trenches for pipework and footings/foundations, *including* operator
Plate compactor	day	Suitable for compaction of base materials under concrete slabs, *excluding* operator
Mobile crane	day	10–15 tonne capacity, diesel or petrol fuel, *including* operator(s)

General items for inclusion in the survey

As noted, several other types of cost contribute to the overall cost of construction. As the ultimate aim is to collect purchaser prices (or their equivalent), additional information is required to allow input costs to be adjusted to represent out-turn costs.

Preliminaries

The term 'Preliminaries' refers to project-specific costs that are not directly a part of the completed project or facility. The term is not universally recognized – there are alternatives like 'general site costs' – but most of the items that typically make up preliminaries are encountered on most projects, for example site accommodation, site fencing and electricity and water supply necessary for construction activities. As noted earlier, the way these costs are broken down and incorporated into project costs may vary from place to place. Preliminaries are often represented as a percentage of net cost (labour+plant+materials), and published price books often give indicative percentages for different kinds of work in different locations. Preliminaries costs vary between downtown, suburban and out-of-town locations, for example, due to differing site access, restricted space for site accommodation and storage and so on. In Australia, for example, these site-specific overheads may account for around 7 to 25+% of the total cost of a project (Rawlinsons, 2010) and must be included if input costs are to be realistically turned into purchaser prices. In the BLOC study (Best, 2008), percentage add-ons for preliminaries for a particular type and location project were gathered from respondents as part of a pricing survey. Given the wide range of percentages suggested in Rawlinsons, some sort of national average percentage is needed. The manner in which these costs are dealt with is likely to vary from place to place; thus a survey question must clearly explain what is to be included. As part of the development of the basket of inputs, the authors produced the following draft question in an attempt to state clearly what was being asked for:

> *What amount, expressed as a percentage of total building cost, would represent a national average for site-specific overheads/costs (sometimes referred to as "Preliminaries")? The amount (percentage) should include all costs incurred in respect of onsite activity including fixed*

cranes, hoists, accommodation, site services, fencing, insurances and the like. It should not include the cost of materials, onsite labour (including onsite supervision), special plant and equipment such as concrete mixers and pumps or general overheads or profit.

Overheads and profit

These two items are often combined and referred to as 'contractors' markup'. General overheads are those costs associated with maintaining the contractor's organization that are not specific to any individual project; typical costs include head office rental/ownership costs and management salaries. Such costs are covered by income from the various projects that the contractor undertakes during a given period (usually a year) and are therefore included in bid prices, and eventually some part of the overall costs become a component of the cost of each project. Profit carries its usual meaning; after all other costs are accounted for, the contractor adds their margin. Markup should not be confused with return on capital, as in many places the cost of construction is covered by borrowed funds, with the contractor investing a relatively small amount of their own capital into projects; hence the survey once again must ask clearly for what is required.

What amount, expressed as a percentage of total building cost (including Preliminaries), would represent a national average for general overheads and contractors' profit (sometimes referred to as "markup")? The amount (percentage) should represent the difference between total net cost of construction (labour+plant+materials+preliminaries) and the contractor's final tender/bid price (excluding VAT/GST).

Tax (VAT/GST, sales tax)

Value-added tax or goods and services tax rates vary considerably around the world. Some countries have no tax of this type (e.g. Hong Kong, United Arab Emirates), while most have some form of tax, with Denmark, Hungary and Sweden (all at 25%) having VAT rates among the highest. Purchasers pay this tax as part of the cost of realizing their building projects. While this data can be readily found for most countries, in some instances the situation is not so simple; for example, in the USA there is no VAT/GST, but there is sales tax on materials, and this varies from zero to around 7% with average of approximately 4.5% (EC Harris, 2009). As in some locations there could be confusion about what tax is applicable, the survey should include a question that permits identification of the amount of tax that is in purchaser prices for buildings. Care is required, as sales tax in the US, for example, may well be included in the prices given for materials, while it is customary in Australia to add GST at the very end of the estimating/tendering process rather than adding to individual inputs along the way.

Consultants' fees

While professional fees paid to various consultants (e.g. architects, engineers) are not strictly part of construction, they are certainly a part of the total cost to the client (purchaser) of creating a built asset. Rawlinsons (2010) suggest total professional fee percentages in Australia ranging from 6.5 to 16% of the total value of the works. It was thought worthwhile that the survey ask for an average percentage for total professional fees across all building types, as this data would then be available for validation purposes even if fees were excluded or dealt with using national statistics.

Conclusion

The foregoing discussion outlines the development of a basket of resources or inputs and some additional items for inclusion for direct input into a model that could produce the equivalent of purchaser prices for building work based on a range of simple input costs as well as provide data for triangulation and validation purposes. The preliminary basket described here was developed further to produce the basket that was eventually used in the 2011 ICP round; details of that basket can be found on the ICP website (see ICP 2011b). Items that were added included some more engineering services items such as a metal storage tank, standby generator and solar collectors. No doubt the basket will be refined further when the data collection exercise from the 2011 round is analysed, and items for which prices were either inconsistent or not provided will be reconsidered.

References

BCIS (2006) *Asia Building Construction Survey* (London: Building Cost Information Service).

Best, R. (2008) *The Development and Testing of a Purchasing Power Parity Method For Comparing Construction Costs Internationally.* PhD thesis, UTS. Available at http://works.bepress.com/rick_best

CIS (2005) *CIS Construction Treatment – 2005.* Paper supplied by V. Kouznetsov.

DLC (2003) *An Initial Test Exercise on a Basket of Goods Approach to Construction Price Comparisons.* Unpublished (London: Davis Langdon Consultancy).

EC Harris (2009) *International Building Costs* (London: EC Harris Built Asset Consultancy).

F+G (2010) *International Construction Intelligence.* Faithful+Gould. www.fgould.com/uk/research-and-features/

G&T (2010) *International Construction Cost Survey.* Gardiner & Theobald. www.gardiner.com/media_publications/economics/?archive

ICEC (n.d.) *The International Cost Model/Location Factor Product Project.* International Cost Engineering Council. www.icoste.org/locationfactor.html

ICP (2011a) *Statistical Capacity Building Activities.* International Comparison Program. http://go.worldbank.org/ONRF1GWBM0

ICP (2011b) *ICP Operational Material, Section B.7 Construction and Civil Engineering.* International Comparison Program. http://go.worldbank.org/4KH0YPSDL0

Rawlinsons (2010) *Australian Construction Handbook.* 28th edition (Perth: Rawlhouse Publishing).

Editorial comment

In 1707, the Bishop of Ely, William Fleetwood, published *Chronicon Preciosum*, generally considered to be the first work devoted to the creation of cost indices. The bishop sought to answer the question of how much £5 would have bought in 1440 compared to what it would have bought at the beginning of the 18th century. The question arose because a scholar at Oxford stood to lose his fellowship, as his income exceeded the maximum of £5 stipulated in a college statute written more than 250 years earlier. By comparing the quantities of some basic goods such as food and cloth which could be purchased with £5 at various times between 1440 and 1707, Fleetwood calculated that £5 in 1440 bought what would cost around £30 in the early 1700s. In effect, the bishop calculated a cost index to make valid comparisons of value over time.

Any cost or price comparisons need data to be normalized, brought to a common basis. A number of chapters in this book review and discuss PPPs for construction that are used instead of exchange rates to adjust for differences in price levels between different countries. In this chapter, Yu and Ive address other important adjustment factors for cost and price differences over time.

The authors review the methods used in the UK for construction input cost indices and output price indices. They note that for more than 40 years, the UK has been a world leader in the field, but they also note that the use of bills of quantities, the basis of the UK's tender price indices, is in decline and that this threatens the future of current methods. The authors also comment on other shortcomings in both data and methods, although the problems are common to most, if not all, countries. They include issues with the collection of data on construction labour costs, an increase in the proportion of work undertaken on existing buildings (the focus of existing methods is on new work), an increase in the proportion of value taken by mechanical and electrical (M&E) services in buildings (excluded from existing methods but now 40% or more of the value of some building types) and difficulties in measuring price changes over time in infrastructure works.

The review is timely; construction methods and construction processes have changed and are still changing, and statistical methods need to keep up to date. Prefabrication, for example, means that more construction work is undertaken off site, and more work on site is assembly of (often bespoke) components rather than handling and fixing individual materials or products. And new forms of procurement such as design and build and public–private partnerships mean that construction work is being procured under very different arrangements than prevailed when current index methodologies were developed. The authors recognize these changes and offer some possible approaches.

There are obvious similarities between temporal cost and price indices and spatial construction price indices (PPPs) and, possibly, efficiencies and economies in producing them using similar data sets and methods. This could be a promising area for future research.

5 A review of construction cost and price indices in Britain

Marco Yu and Graham Ive

Introduction

What determines the living standards of a society is the quantity of goods and services produced by the society. The importance of measuring this quantity is obvious for the understanding of economic progress. However, the statistical agencies of governments measure the monetary (nominal) value of the goods and services produced by their countries (usually called GDP at current prices). This value is a set of (not directly known) quantities multiplied by a set of (not directly known) prices. The conversion of the monetary value of output to the real output of the economy requires a price index because the changes in monetary value of the goods and services produced are the combination of two movements: monetary price level movements and quantity movements. Therefore the measure of the real outputs is only as accurate as are the price indices in measuring price changes, and hence the construction price and cost indices in construction productivity research in which the focus is on understanding the relationships between the changes in real inputs and outputs of the construction industry over time are of importance.

Our initial reasons for questioning the quality and accuracy of the existing British construction price and cost indices are threefold. First, Allen (1985, 1989), Dyer *et al.* (2012), Goodrum *et al.* (2002), Gordon (1968), Ive *et al.* (2004) and Pieper (1989, 1991) consider the biases in the published construction price and cost indices in, variously, the UK, US, France and Germany, as one of the probable main causes of inaccurate measures of productivity growth rates in the construction sector.

Second, the existing compilation method of British construction price indices was developed in the late 1960s and that for construction cost indices in the 1970s, and since then, there have been profound changes in construction technologies, changes in procurement routes and the associated contract documents, as well as the growing significance of mechanical and electrical services. For example, building projects procured via design and build, which has been gaining in popularity, are completely ignored in the existing building price indices.

Third, advances in the general economic theory of indexing have not been incorporated in the methods used for British construction price indexation since the late 1960s. Recent improvements such as hedonic price indices have been adopted by the UK, US and Germany statistical agencies, to name but a few, to compile other price indices.

Most research on construction price and cost indices is about forecasting and modelling using time series or other techniques, in which past values of price indices or other variables are used to forecast future values of the indices and thus construction price and cost inflation (Akintoye and Skitmore 1994; Akintoye *et al.* 1998; Fellows 1991; Hwang 2009, 2011; McCaffer *et al.* 1983 and Ng *et al.* 2004). Such models presume that the published indices do accurately measure construction inflation. To fill this 'presumption' gap, this chapter aims to scrutinize the compilation method used for construction price and cost indices with a view to indicating what is and what is not actually being measured, and identifying opportunities for improvements.

The construction sector is complex and its projects are heterogeneous, including infrastructure as well as buildings, and work to existing structures as well as new projects. Therefore, we require a number of price and cost indices measuring the inflation in each subsector of the construction industry (both new work and work to existing structures). As shown in Table 5.1, only about 50% of the output of the construction sector is believed to comprise projects of the kind covered by the tender price indices. The remaining 50% not covered consists mainly of repair and maintenance (40%) and new infrastructure work (10%). Buildings account for more than 80% of the new work output and infrastructure for the remainder. In comparison with new infrastructure and all repair and maintenance work, the output of the new building subsector is less diverse and easier to measure, and as a result, the price indices of new building work are relatively well developed. Moreover, new building work in Britain, as will be shown, has traditionally been the field of application of bills of quantities (BQs), in which the aggregate values of successful tenders are broken down into unit prices for specified quantities of elements in the finished building. The British method of compilation of tender price indices is based upon this fact. BQs, of course, only exist for new building and some major refurbishment projects and not for repair and maintenance projects. Moreover, it is noteworthy that the samples collected for compiling the British tender price indices (TPI) do not even represent the subsectors within new building work proportionally. We return to this later.

TPIs represent attempts to measure the inflation of the contract prices between clients and contractors for constructing 'new' buildings, which includes some alteration and improvement work to nonresidential buildings. In addition, TPIs are components of the deflators (output price indices) used to derive the building industry's real new output (output at constant prices). Construction cost indices (CCIs) are attempts to measure the inflation of the input prices (such as labour wages, construction material prices and plant hire prices)

Table 5.1 Construction Output of Great Britain at Current Prices, 1983 to 2012
(Source: ONS Statistical Bulletin: Output in the Construction Industry, various issues)

Year	All work (£ million)	New Work		Repair & Maintenance
		Building	Infrastructure	
1983	26,623	52%	11%	38%
1984	28,990	51%	10%	39%
1985	30,957	51%	9%	39%
1986	33,427	52%	9%	39%
1987	39,143	53%	8%	38%
1988	46,811	55%	9%	36%
1989	54,391	56%	9%	35%
1990	57,673	54%	10%	36%
1991	53,378	49%	14%	37%
1992	49,602	46%	15%	39%
1993	48,431	45%	15%	40%
1994	51,647	47%	13%	40%
1995	55,013	46%	13%	40%
1996	57,749	45%	14%	40%
1997	60,990	47%	13%	40%
1998	64,829	49%	12%	39%
1999	68,582	52%	11%	37%
2000	72,714	52%	11%	37%
2001	77,958	51%	11%	38%
2002	87,218	52%	12%	37%
2003	97,259	53%	10%	37%
2004	106,657	57%	8%	36%
2005	111,494	57%	7%	36%
2006	118,320	58%	7%	35%
2007	127,062	59%	7%	34%
2008	128,641	56%	8%	37%
2009	111,079	51%	10%	39%
2010	117,385	52%	12%	36%
2011	121,737	51%	13%	36%
2012	115,583	49%	12%	39%
	Geometric Mean	52%	10%	38%

to contractors. The UK Department of Business, Innovation and Skills (BIS) (BIS 2013) surveyed the main uses of their published construction price and cost indices. The four most important uses of the TPI and CCI are as follows:

- deflation of construction-sector components of the nominal national product to produce estimates of real output from the sector
- capturing relative price change and inflation in the construction industry for assessments and forecasting of market conditions

- updating historical cost data for cost planning and estimating
- international, intersectoral or intertemporal comparisons of the level and growth of price, real output and productivity.

Figure 5.1 presents the trends of the Building Cost Information Service (BCIS) All-in Tender Price Index and the BCIS General Building Cost Index. This example clearly shows that they deviate in terms of levels, growth rates and volatility, and therefore, even assuming both to be accurate, CCI would not be a good proxy for the TPI. If similar relationships hold elsewhere, this undermines the validity of using CCI as the deflator of construction output as is done in some countries.

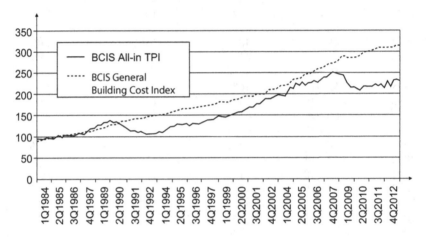

Figure 5.1 BCIS All-in Tender Price Index (TPI) vs. BCIS General Building Cost Index (CCI) (BCIS On-line)

The next section describes the current TPI and CCI compilation methods generally adopted in Britain and their development and evolution. The following section provides an evaluation of the fitness for purpose of the TPI and CCI compilation methods and suggests the most important areas to be addressed. In the section after that, possible ways to improve the TPI and CCI are discussed.

Methods used to compile tender price indices in Britain

BIS produces the most extensive public-sector TPI. The BCIS, the building information research arm of the Royal Institution of Chartered Surveyors (RICS), compiles its own building TPI drawing on its wide reach to private and public projects through the willingness of RICS members to supply data to their own chartered professional institute. Davis Langdon (DL, now part of AECOM), one of the largest quantity surveying practices, also publishes its own tender price index. However diverse the sources of information that these three organizations utilize and however differently the resulting indices signal market conditions, the TPI compilation methods behind their array of indices are very similar, and the origin of the method can be traced back to a joint task force of representatives of the RICS, University College London (UCL) and the then Ministry of Public Building and Works (MPBW).

Under the auspices of the MPBW and the RICS, Bowley and Corlett of UCL produced a report on trends in building prices (Bowley and Corlett 1970). The building price indices being published at that time by the government were input cost indices for labour and material cost, which would from time to time differ from the trends of tender prices in the building industry due primarily to changing productivity and/or market conditions (Fleming 1965). It was against this background that Bowley and Corlett reviewed several possible methods to compose a true tender price index for the building industry, and the method described in Chapter 5 of their report became the workhorse method used in the then Department of the Environment (DoE, now BIS), BCIS and DL ever since.

The method used in BIS is described first and then differences in the methods adopted in BCIS and DL later are highlighted. Mitchell (1971) made the first attempt to document the method used in the then DoE. The following description mainly relies on a manual produced for internal use by the then Quantity Surveyors Services Division (QSSD).

First of all, the data BIS collects for compiling their public nonhousing tender price index is the accepted BQs of building projects procured in a quarter of a year (known as the reference period). Under the traditional procurement route, the client of a building project employs quantity surveyors to quantify the building work as much as possible from the design, which facilitates the preparation of bids by construction firms based on a common framework. The bills of quantities comprise a number of bills, and each bill traditionally covers a separate trade or element. The bill items capture the quantity in suitable physical units, such as cubic metres, of, for example, in situ concrete to be contained in the finished building, as 'taken off' the drawings prepared by the design consultants. The construction firms compete by attaching different prices to each unit of measured work.

In addition to the measured work, there is typically a section called Prime Costs and Provisional Sums. Prime costs are usually allowed for specialist work (not designed by the architect) such as lifts, heating systems, air-conditioning systems and electricity supply systems, whereas provisional sums are for the work for which the design is not detailed enough to allow quantification, for example, landscaping. Therefore, works allowed in the Prime Costs and Provisional Sums section will be adjusted in the future according to the actual cost incurred, and the construction firms compete on the markup on these works, which is supposed to cover their profit and their overhead expense incurred because of these works. Traditionally, there is also a section of BQs called Preliminaries which covers the contractors' general costs for executing the work as a whole. Therefore, the tender price is the summation of the bill items, prime costs and provisional sums, the preliminaries, and other adjustments such as commercial discounts.

BQs provide a rich source of information about the prices and quantities of various elements of the measured work of building projects at the reference period. To construct a price index, the prices at the base period (here, 1995) are also needed. When BIS analyse a BQ, they will reprice it by the rates in the Property Services Agency (PSA) Schedule of Rates of the base year, supplemented with some BQ rates they have collected at the base year. The BIS Public Sector Building (Non-Housing) TPI is a fixed-base, matched-item Paasche index.

To produce an index for each project (project index), from each trade of the project, the items are repriced in a descending order of value until the repriced items are equal to more than 25% of the value of the trade and all items with values greater than 1% of the measured work total are repriced. Therefore, it is a current-weight Paasche index. As only items that can be matched will be compared, so it is a matched-item index.

The sum of all items repriced at the rates of the Schedule of Rates is divided by the sum of the corresponding values at the bill rates with the allocated adjustments on measured work in the BQ to obtain a Schedule Factor.

Schedule Factor:

$$= \frac{\text{Sum of the selected items being repriced at the Schedule of Rates}}{\text{Sum of the selected items at the bill rates } + \text{ allocated adjustments on measured work}}$$

The adjustments on measured work are the adjustments made on the main summary of the BQs such as head office overhead, correction of arithmetic errors and commercial discount. These adjustments are allocated to the selected items *pro rata* according to their values.

With the Schedule Factor, the project index is computed by this formula:

Project Index:

$$= \frac{\text{Contract Sum less Dayworks and Contingencies}}{\text{Contract Sum less Preliminaries, Dayworks and Contingencies } \times \text{ Schedule Factor}}$$

The reason for deducting the preliminaries from the contract sum in the denominator is that the rates in the Schedule of Rates include allocated preliminaries.

Since *location* and *function* of the building are believed to be main cost drivers, and BIS want to reflect the general building price over time, independent of changes in these factors, each project index number is adjusted for these factors. The published index number is then the median value of these adjusted project index numbers in the quarter and is smoothed by use of a three-quarter moving average.

It is a fixed-base index because the Paasche index is a bilateral index. To construct a time series price index, BIS choose the same base year, say 1995, to compare all the subsequent BQ rates. Therefore, all the later year indices are compared against the 1995 Schedule of Rates. BIS have from time to time changed the base Schedule of Rates. In the past, the base Schedule of Rates was changed every five years, but rebasing has become less frequent, and the latest PSA Schedule of Rates produced by Carillion (one of the UK's largest construction contractors) was rebased in 2005 prices.

The BCIS index is also a fixed-base, matched-item Paasche index. It matches comparable items and uses the current quantities in the BQs to weight the prices. BCIS use the same Schedule of Rates for the base prices as BIS. They only sample projects of more than £100,000. The difference between the BCIS and BIS methods lies in the way they adjust and aggregate project indices. From 1984, each BCIS project index has been adjusted for the *size*, *location* and *contract type* (*firm price* or *fluctuating price*) before aggregating into the published indices (whereas BIS adjust for *location* and *function* only). The other salient difference is that BCIS takes the geometric mean rather than the median of the adjusted project indices. The BQs, which cover both public and private sectors, are supplied by RICS members.

The Davis Langdon (DL) TPI is a chain-linked, matched-item Paasche index. The obvious difference is the application of the chain-linked system to join up the bilateral indices. In a chain-linked system, the reference period of the previous bilateral index becomes the base period of the succeeding bilateral index. For example, if 2012Q4 is the base period and 2013Q1 is the reference period in the first quarter, then in the second quarter, the base period is 2013Q1 and the reference period is 2013Q2. As DL publish a price book – *Spon's Architects' and Builders' Price Book* – annually, the base prices are actually updated every year rather than quarterly. The index is, therefore, more accurately called an annually chain-linked index, and as such is similar to the Consumer Price Index (CPI) and Retail Price Index (RPI) compiled by the Office for National Statistics (ONS). Since the sample is confined to the projects in which DL is involved in Britain, and the sample size in number of projects is therefore smaller, more than 25% of items in terms of value are sampled in each project to reduce the sampling errors. The adjustment factors of the project indices are *size*, *location* and *building function* and the geometric mean is used to aggregate the project indices. The two advantages of the method used by DL are:

- it is chain-linked not fixed-base
- there are three adjustment factors not two, but its disadvantage is that DL draws on a smaller and less representative sample of BQs.

Compilation methods for construction cost indices in Britain

BCIS, BIS and DL publish general construction cost indices to reflect the inflation of the input prices paid by contractors in various subsectors of the construction industry. These cost indices are developed from the Price Adjustment Formulae indices, also known as NEDO indices (originally compiled for the Construction Committee of the National Economic Development Office).

Construction contracts with fluctuation provisions allow contractors to pass on to their clients the increase in input costs, such as wages and material prices, in the period between the date of tender and the work being carried out. NEDO's Steering Group on Price Fluctuations Formulae published a report in 1969 which proposed a formula-based method to calculate input price (i.e. cost) fluctuations in building contracts. It suggested dividing the contract sum into trade-based work categories (such as brickwork and concrete) and adjusting by the published indices for each work category. By analysing 60 completed questionnaires, the report concluded that the administrative cost of agreeing the fluctuations by the recommended formula method would be 0.16% of the contract sum compared to 0.75% by the conventional method of auditing suppliers' invoices and the wages set by the appropriate wage-fixing bodies.

The formulae methods of adjusting fluctuations in civil engineering contracts began in 1973 and in building and specialist engineering contracts in 1974. Each work category index is a weighted index of various labour wages, material prices and plant costs that reflect the cost of that particular work category. No productivity growth is assumed in the work category indices.

The sources of the labour wages are largely based on national labour agreements. At the time, quite a large proportion of the construction workforce had wages based at least in part on rates set by the national wage agreement bodies, which had representatives from employers and unions. This is no longer the case.

The construction material price indices are compiled by ONS as part of the whole-economy producer price indices. BIS publishes the material price indices in its *Monthly Statistics of Building Materials and Components*. The plant cost, including the cost of the operator, is a weighted average of depreciation, building labour cost and consumables such as tyres and fuel cost.

The weightings of labour wages, material prices and plant cost for the work categories have been revised infrequently since the first series was published in 1974. The second version, Series 2, was published in 1977, and the latest version, 1990 Series (also known as Series 3), was published in 1995.

The double-digit inflation that plagued the UK for the majority of the 1970s and 1980s made the fluctuation construction contracts very popular, and the Price Adjustment Formulae indices then had an important role to play. Inflation in the UK has come down under 5% since the mid-1990s and, consequently, fluctuation contracts have become exceptions rather than the norm.

BCIS publish nine building cost indices on a monthly basis:

- general building cost index
- general building cost, excluding mechanical and electrical (M&E), index
- steel-framed construction cost index
- concrete-framed construction cost index
- brick construction cost index
- mechanical and electrical engineering cost index
- basic labour cost index
- basic materials cost index
- basic plant cost index.

BCIS (1997) explained that it analysed 54 bills of quantities to work out different weightings for the various work category indices for each of the building cost indices. Since the weightings are fixed in the base year (1977), the BCIS Building Cost Indices are fixed-base matched-item Laspeyres indices. Figure 5.2 uses the BCIS Brick Construction Cost Index as an example to illustrate the relationship with the work categories indices and the underlying input indices.

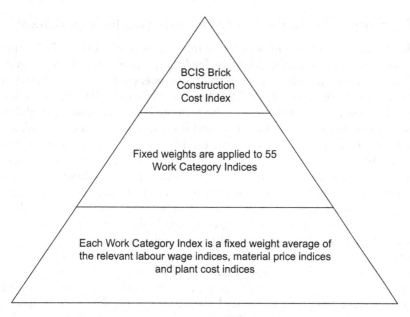

Figure 5.2 Relationships among construction cost indices, work categories indices and underlying input indices

Table 5.2 Construction Resource Cost Indices published by BIS

		Combined Labour & Plant	Materials	Mechanical Work	Electrical Work	Building Work
New	Building Non-housing		Available			
	House Building		Available			
	Road Construction	Available		Not Applicable		
	Infrastructure	Available		Not Applicable		
Main-tenance	Building Non-housing		Available			
	House Building		Available			

BIS publish 35 resource cost indices for construction in the UK on a quarterly basis covering repair and maintenance as well as new work. Table 5.2 summarizes the availabilities of the cost indices in the different sectors and for the different inputs of the construction industry.

BIS (2012) reported that the weightings of the new building nonhousing index had been assessed in the 1970s, and the weightings of the rest were assessed by a panel of Chartered Quantity Surveyors in 1998.

Evaluation of the British TPI and opportunities for improvements

Before making any recommendations for improving the British TPI compilation method, it must be acknowledged that, having attempted to review the many different compilation methods documented for other countries, as summarized, for example, in OECD and EUROSTAT (2001) and EUROSTAT (1996), it is clear that the method adopted in the aforementioned three British organizations has led the world for a quarter of a century. Blessed with the availability of BQs, the British TPI does measure the output price of the building industry by making use of the contract prices as opposed to many other countries that use input prices such as labour wages and material prices as proxies for output prices. For instance, until recent years, the US agencies used input prices for deflating the output of the building industry, a procedure that has long been criticized in the United States (Gordon 1968; Pieper 1989, 1991). Although criticisms have finally led to the introduction in the US of a new building output price index by the Bureau of Labor Statistics and US Census Bureau, its prices are either deduced from the property price by attempting to remove the land value from the former or from questionnaires, and as such the validity is less than that of the contract price data obtained from BQs in Britain.

Despite its many advantages over systems in use elsewhere, the following opportunities for improvements in the British system have been identified.

Mechanical and electrical service items

Except for plumbing work, all mechanical and electrical (M&E) service items including comfort cooling, heating systems, lighting, electrical supply systems, lifts and fire-detection systems are not measured in the index because mechanical and electrical services are usually included as prime costs or provisional sums in BQs. In some nonresidential buildings such as offices and hospitals, mechanical and electrical services represent a significant portion, approximately 40%, of the total cost of the building. Leaving this out could result in significant measurement errors.

Figure 5.3 shows that during the 1980s, the building cost index and the mechanical and electrical cost index tracked each other closely, but during the 1990s, the mechanical and electrical cost index was consistently at a higher level than the non–mechanical-and-electrical building cost index. From around 2005 the trend reversed. This reflects the fact that the mechanical and electrical services and building work input markets are subject to different short-run cost drivers. If we look at the weightings of the cost indices, material prices have a higher weighting in the M&E cost index.

As previously noted, the building and M&E cost indices assume no productivity growth and are fixed-weight averages of the producer price indices of materials and of the wages in national labour agreements. The weightings were obtained from analyses of 54 bills of quantities for new building work (BCIS 1997).

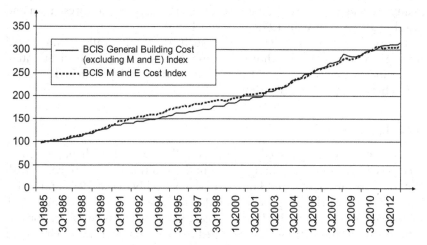

Figure 5.3 BCIS Building Cost and M&E Cost Indices (1985 = 100) (BCIS Online)

It is generally observed that goods and services from a sector with higher technological progress and productivity growth have lower price inflation than those from a lower productivity growth sector. The personal computer is a typical example of the former, whereas haircutting services is a widely cited example of the latter.

Mechanical and electrical services, such as air conditioning and heating systems, are reckoned to have been subject to higher productivity growth in the past than the more traditional building trades, such as brickwork, being measured by the TPI. Anecdotal evidence in regard to ICT cabling also suggests that the quality of the cable has increased, say from cat 5 to cat 6, but the nominal prices have been stagnant or have even fallen. Another example is the significant drop in the supply and installation price for domestic solar PV since 2010 in the UK.

New elements and proprietary items

Since the method is to compare the prices of BQ items with the prices in the base schedule of rates, the price of new goods or proprietary items that cannot be matched will not be measured in the TPI. For new goods, frequently updating the base schedule of rates will alleviate part of the problem, and that is the reason ONS adopts an annually chain-linked system for compiling the RPI and CPI. In other (fixed-base) methods, the effect of introduction of new goods will not be measured, and ignoring this will often result in an upward bias of the price index because new goods can usually achieve the same outcome at a lower price than the old goods being replaced. Nordhaus (1997) demonstrated, for example, that ignoring the introduction of new goods substantially overestimates the true price of lighting over time.

Despite its importance, the appropriate method for estimating the price change of a new good when it is introduced is controversial (Bresnahan and Gordon 1996; Hausman 1999; Wolfson 1999).

The problems of proprietary items such as curtain walling and glazed internal partitions are also thorny because the design of the proprietary item is specific to each project, and this prevents them being matched or compared between projects over time.

Sample coverage

RICS (2006, 2012) revealed a clear overall shift of British procurement methods from lump-sum design-bid-build (traditional procurement) with BQ to lump-sum design and build over the period between 1985 and 2010. The share of workload procured under lump-sum design and build has increased from 8.0% to 39.2% by project value, whereas the share for traditional lump sum with BQ has dropped from 59.3% to 18.8%. This trend is unlikely to reverse because the design-and-build procurement route is widely adopted in private commercial projects and private finance initiative schemes and their variants. However, BIS, BCIS and DL only survey the BQs of the traditional procurement method for their TPI calculation. With the dwindling popularity of the design-bid-build with BQ method, continuing to rely on BQs for compiling TPI would make TPI prone to larger sampling errors or even biases. Emphasis needs to be placed on measuring the price movements in design and build contracts.

Sample size and distribution

BCIS aims to sample 80 projects in each quarter because it believes that if 80 projects are sampled, about 90% of the price indices of individual projects will cluster within a reasonable region (about ± 2.8 %) around the average. In the period between 1990 and 2012, the BCIS All-In TPI had an average quarterly sample size of 63, of which 36 were public-sector, non-housing building projects. By contrast, BIS sampled 57 public nonhousing building projects on average in each quarter over the same period for its Pubsec TPI. Since the index compilation method adopted in BCIS and BIS is similar and both BCIS Public TPI and BIS Pubsec TPI measure the inflation of tender prices in the same domain, there is room for collaboration and specialization.

It would be advisable for BCIS to focus its effort and resources on collecting private-sector information, thereby increasing the sample size of private-sector projects. Currently two subindices of the BCIS All-In TPI, namely BCIS Private Commercial TPI and BCIS Private Industrial TPI, serve as data that BIS use to construct the construction output deflator because these two indices capture the tender price movement of the private sector, to which BIS has no access.

As mentioned in the introduction, Figure 5.4 shows that the distribution of BCIS samples over the period 1990 through 2012 is not aligned with the percentages of output of new building work. Of the 31% of all new output that is in the housing sector, private work accounts for 26.7%, while public work accounts for 4.6%. BCIS, however, note that the majority of their housing samples come from social housing projects. Therefore, the public housing sector is overrepresented, but the construction price movement in the private housing sector is hardly measured in the TPI. Since speculative builders in the private housing market may perform the dual role of developer and main contractor, the tender prices of the construction work, let alone bills of quantities, are generally not available. This problem is not specific to Britain; for example, the US Census Bureau estimates the value of construction work in the private housing market by applying a fixed ratio (currently 84.24%) to the average sales prices of houses.

Private commercial work also appears to be underrepresented in the samples. A significant portion of the private commercial work is major refurbishment of existing buildings which are not measured in the BCIS All-In

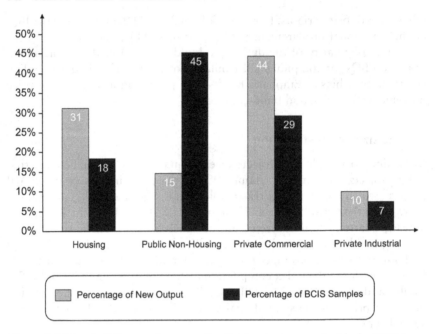

Figure 5.4 New building work output distribution is compared with the BCIS Sample distribution, 1990 to 2012. (Source: Authors' calculation, BCIS Online and ONS Statistical Bulletin: Output in the Construction Industry, various issues)

TPI. However, using the same methodology, BCIS introduced a refurbishment TPI in 1991 with a sample size of around 14 per quarter, so that potentially the All-In TPI samples could be extended to include commercial major refurbishment.

It is also noteworthy that since its sample size of private industrial projects has become too small, BCIS has adopted a different method to compile the TPI for the sector since 2010. In brief, it makes use of the trade price information collected in the BQs of other types of projects and reweights them using the historic BQs of private industrial projects.

Two possible ways to move forward

Having reviewed and assessed the existing TPI compilation methods, this section considers some ways to improve them. Even with suggested changes (see below) in the existing matched-item Paasche indices to improve coverage and samples and to update them to annually chain-linked, it is difficult to cater for the price movements of the diverse M&E items and the effect of quality changes on prices in the existing method. Against this background, we propose the employment of hedonic techniques as a supplement.

Improving the existing method

Regarding the sources of price information, the diminishing popularity of the traditional design-bid-build with BQ procurement route is a real challenge. Some design-and-build contractors produce full BQs for bidding or cost management purposes, and it is worthwhile pursuing the accessibility and pervasiveness of such information.

Alternatively, the feasibility of using cost plans in the BCIS Standard List of Building Elements format deserves further research. The majority of design-and-build projects in the PFI market and private sector include cost plans in the contract documents, and the rates in such cost plans are in principle comparable to schedules of rates such as those in the Approximate Estimates section of *Spon's Architects' and Builders' Price Book 2013*.

This cost information may be less reliable than BQs for reflecting the true prices of various components of buildings, but it is still better than totally ignoring this growing sector. Also, the quantities measured in those cost plans are useful input information for a hedonic index, something discussed in the next section. There is potential for a circular relationship because the TPIs may be used in setting the cost plan rates, but measures are taken in practice that mitigate this concern: contractors market-test the significant cost elements before submitting a firm price bid, and professional QS firms working for clients ensure the prices in cost plans reflect market prices.

The current method only compares prices of items accounting for 25% of each measured trade by value. Mitchell (1971), Azzaro (1976) and BCIS (1983) show that the 25% rule was a practical compromise between stability of the index and the production cost given the computer technology of the early 1980s. Mitchell reports that the number of items to be compared for 25%, 50%, 75% and 100% of the trade value are 40, 98, 175 and more than 1,000, respectively. Mitchell found the project indices of 80 BQs using the 25% rule to be as stable as that using full repricing (100% rule). However, the 25% rule produced an aggregate index 4.4% higher than the full repricing index reported by Mitchell (1971), whereas the 25% rule underestimated the full repricing index by 1.1% in a separate study reported by Azzaro (1976). There is a case for repeating these studies with more recent data. If this shows discrepancies in estimated levels, then, with the advance of computer technology over the last quarter of the century, it is practical, at least for public-sector projects, to extend the sample items to far more than 25% by value.

DL has adopted the annually chain-linked system, which allows them to compare rates of new items more quickly than the base-linked system in BCIS and BIS with their less-frequent revisions of the base schedule of rates. As early as 1887, Alfred Marshall suggested that the chain-linked system would be a better measure of the price impact of invention of new commodities. The main difficulty in converting the current indices in BCIS and BIS to annually chain-linked indices is the need to update their base schedules of rates annually. RICS acquired a well-established building price book publisher in 2005 and merged it with BCIS, enabling, in theory, BCIS to convert its TPI to an annually chain-linked system by using their building price book published annually rather than the dated PSA Schedule of Rates for the base period rates. The methodology for compiling the PSA Schedules of Rates and the building price book is much the same. With the many annually published UK–based building price books such as BCIS Wessex, Griffiths, Hutchins and Laxton's and the similarity of the methodology between these price books and that of the PSA Schedule of Rates, it is feasible that BIS could also switch to an annually chain-linked system for their TPI.

Hedonic price index

The hedonic regression technique has been gaining acceptance among statistical agencies such as ONS in the UK and US Census Bureau for compiling their price indices. Ball and Allen (2003) reported that the statistical agencies in Canada, Finland, France, Germany, Sweden and the US have used the hedonic technique to adjust for quality changes in electrical goods such as personal computers, dishwashers and TVs in their price indices. ONS have used the technique to adjust for quality improvements in personal computers, digital cameras, laptops and mobile telephone handsets since 2003 (Fenwick and Wingfield 2005). In real estate and construction statistics, the US Census Bureau has used the hedonic technique to produce their single-family house construction price deflator since 1968 (Musgrave 1969), and ONS have applied it to estimations of imputed rents for owner-occupiers' houses (Richardson and Dolling 2005). In both cases, hedonic regression techniques are used to adjust the heterogeneities among buildings rather than adjust for improvement in quality over time. Meikle (2001) suggested hedonic methods as an area for further research on construction price indices.

What is a hedonic function? People value a good for its attributes or characteristics. Therefore, goods can be regarded as bundles of attributes, and their values are the sums of the values of the attributes within the bundles. Hedonic function refers to the relationship between the price of the good and the implicit prices of the various attributes embodied in the good. If quantities of attributes are measurable, regression techniques are commonly used to estimate the hedonic function of the good from historical data.

One promising application of hedonic price indices is to extend the coverage to projects that do not have BQs. Figure 5.3 suggests that the current BQ–based TPIs would be unrepresentative for the private housing, private commercial and private industrial sectors. Although hedonic indices have been applied to single-family housing in the US and lessons can be learnt from the relevant research (e.g. Somerville 1999; Dyer *et al.* 2012), it is probably not the most rewarding sector for the application of hedonic price indices in the UK because of the difficulty of separating the construction price from the total sale price.

A growing number of studies do, however, estimate the relationships between various attributes of buildings and their construction prices. Emsley *et al.* (2002) and Lowe *et al.* (2006a, 2006b and 2007) have identified some attributes driving construction price in the UK. Table 5.3 summarizes the significant price-driving attributes reported by some of the studies.

Table 5.3 Construction price-driving attributes from selected literature

Research	Data	Attributes
Thalmann (1998)	15 residential projects in Switzerland	• Total useable floor area • Proportion of openings in external wall • Proportion of external walls that lie underground
Elhag and Boussabaine (1999)	36 office buildings	• Gross floor area • Project duration
Kim *et al.* (2004)	530 residential projects in Korea	• Gross floor area • Stores • Units • Project duration • Roof types • Foundation types • Usage of basement • Finishing grades
Chan and Park (2005)	87 projects in Singapore covering residential, industrial, offices and schools	• Function of the buildings
Chen and Huang (2006)	132 school reconstruction projects in Taiwan	• Floor area • Project duration
Emsley *et al.* (2002) Lowe *et al.* (2006a, 2006b, 2007)	286 projects in the UK covering industrial, commercial, educational, health, recreational, religious and residential	• Gross internal floor area • Function • Project duration • Mechanical installations • Piling or not
Stoy and Schalcher (2007)	290 residential projects in Germany	• Gross floor area • Median floor height • Share of the ancillary areas for services • Project duration • Compactness of the building
Blackman and Picken (2010)	36 residential buildings in Shanghai	• Gross floor areas • Height of the buildings
Ji *et al.* (2010)	124 apartment buildings in Korea	• Gross floor areas • Number of apartment units • Number of floors

To construct a hedonic price index, the dependent variable of the model is the price that the clients pay for the construction of the buildings. Hedonic functions of construction price for base period and reference period are estimated respectively. By inputting the attributes of a building built at base period into the hedonic function for the reference period, the reference period construction price of the building can be estimated. A Laspeyres index can be constructed by comparing the derived reference period prices to the base period prices. It is obvious that a Paasche index can be constructed by using the attributes of reference period buildings and hedonic functions; a Fisher ideal index can then be constructed from the Laspeyres and Paasche indices. Triplett (2004) provides a few other alternatives for the construction of a hedonic price index.

The price-driving attributes identified in the literature are mainly used to adjust the heterogeneities among buildings, such as the floor areas and functions of the buildings, rather than to adjust for aggregate average quality improvement over time. The BIS Tender Price Index (TPISH) published by BIS has also applied the hedonic technique to control the variations in project specifications.

Quality adjustment is more important in the hedonic price index for computers (Cole *et al.* 1986; Pakes 2003; Silver and Heravi 2004). The common attributes in the hedonic functions include the speed of the CPU and the memory of the hard disk. These are 'vertical' attributes since consumers prefer more of them (faster CPU and 'larger' hard disk) rather than less. They capture the quality improvement of computers over time.

A definitive solution for measuring inflation in the prices of mechanical and electrical services in buildings over time is elusive. However, the hedonic regression technique can shed light on this, and the following offers some suggestions for further research. Performance specifications for mechanical and electrical service systems are usually produced by professional engineers, appointed by clients. The first task is to translate these performance specifications into measurable input and output attributes of the systems. Prominent resource and input cost heterogeneities between systems such as underfloor heating systems versus traditional profiled surface radiator systems should be included in the hedonic model as dummy variables. Focus, however, needs to be given to the vertical (output, performance) attributes and capacities of the system, and energy efficiencies of the system could be two main attributes to be captured in a hedonic model. Capacities refer to the maximum power (kVA) of electricity generators, maximum loading of lifts and so on. The total or net area served by an M&E system is a good example of the capacity of a system. Energy efficiency is the unit of effective output of the system per energy input. The Seasonal Efficiency of Domestic Boilers in the UK (SEDBUK) for gas, LPG or oil boilers and luminaire-lumens per circuit-Watt of the lighting system are two examples of measures of energy efficiency of mechanical and electrical services. When Ohta (1975) produced a quality-adjusted price index for the US boiler and turbogenerator industries, he applied a hedonic technique to cater for the efficiency and capacity improvement of the boilers and turbogenerators over time. Berry *et al.* (1995) also found that the capacity variable (horsepower per unit weight) and efficiency variable (miles per dollar) played a key role in their hedonic model of automobile prices.

After measuring the attributes of mechanical and electrical service systems, the next task is to collect price information for the system. For building projects being procured via the traditional route, the sum can be found in the Prime Cost section of the BQ. The prime cost sums are fairly accurate since they usually reflect the fixed prices agreed between clients and nominated subcontractors. With the adoption of new standard forms of contract such as JCT 2005, nominating subcontractors has become less popular, and now the usual arrangements for procurement of M&E and other specialist trades in the traditional route are via Contractor's Design Portion. Contract sum analysis of the M&E services is usually provided, which provides useful information for hedonic analysis. In design-and-build procurement, the mechanical and electrical service costs normally become part of the fixed price lump sum of the contract and can be discerned in the cost plan of the contract documents.

A hedonic index of mechanical and electrical services, if adequately developed, will be a significant supplement to the existing TPI method since it will capture the price movement of the most cost significant component of buildings unmeasured by the existing method. Perhaps, with the richness of tender price information, a hedonic index of non–M&E tender prices could also be developed. If so, its performance could then be monitored against the TPI compiled by the existing method. The challenge would be to develop performance measures for non–M&E elements of buildings as relevant and potentially precisely measurable on a continuous scale as the performance measures developed for M&E services. It is, however, encouraging to note that performance specification has grown in popularity in the US infrastructure construction sector (Guo *et al.* 2005), and performance-based contracting has been proposed in the UK (Gruneberg 2007).

A school of thought in the industry is that the rates in BQs are distorted by front-end loading strategies, opportunistic bidding behaviour where low rates are applied to small-quantity items, idiosyncratic methods for allocation of preliminaries, overhead and profit in BQs and so on. If these problems are important, one of the advantages of a hedonic TPI over the traditional TPI would be that it does not rely on the rates in BQs but on market prices of subcontracts. Moreover, factors such as locations, sizes and functions of buildings that BIS, BCIS and DL adjust for in the TPI compilation process could, with a hedonic index, be explicitly modelled.

Evaluation of the UK construction cost indices and opportunities for improvements

Weightings

The construction cost indices reviewed earlier are fixed-base, matched-item Laspeyres indices. Since the method uses the weighting fixed in the base year, it does not allow for substituting cheaper inputs for more expensive inputs in the reference year and would tend to overstate inflation or understate deflation. This problem can be alleviated by updating the weighting more frequently. However, the last update of weightings of the Price Adjustment Formulae was in 1995.

Quality of the input and productivity growth

It is important to emphasize that construction cost indices are not intended to reflect reduction in cost due to productivity growth in the construction sector. In brief, productivity growth means requiring less input to produce the same output. When contractors find a way to use fewer man hours or less material to produce the same output, their 'input cost' should drop, but the current compilation method used for the construction cost index would not capture it. A related but different issue is the quality of the input. Both the quantity and quality of the input are multidimensional, and the input price can only be based on one dimension of the quantity. For example, a wage is usually based on time, while concrete and steel are based on volume and weight. The other dimensions of the quality of input, such as the skill level of the labour and the strength and durability of steel, are assumed to be constant, which may not be the case and would bias the cost indices.

Material price

A common criticism of the construction cost index is that the material prices are 'list prices' and discounts are ignored. However, ONS reports that it does attempt to collect real transaction prices when compiling their producer price indices, and it is outside the scope here to verify this. The UK Statistics Authority (2011b) reviewed methods of the *Monthly Statistics of Building Materials and Components* produced by BIS and confirmed their status as National Statistics.

Labour wage

The labour wage component of construction cost indices based on the national labour agreements is more concerning. There could be a variable time lag between market conditions and the wages reflected in the national labour agreement. In addition, given the change in unionization in the UK construction industry over time, it is likely that such national wage agreements will deviate from market wages.

Table 5.4 and Figure 5.5 compare the labour cost indices based on national wage agreements and the survey-based labour cost collected by ONS.

The national labour wage agreement indices (BCIS Labour Cost Index covering buildings and Civil Engineering Labour Cost covering civil engineering) show a higher growth (more than 70% between 2000 and 2012) than other measures of construction labour cost or earning indices (*circa* 50%) reported by various ONS surveys.

While a detailed examination of various labour earnings indices is outside the scope here, it is worthwhile to compare these indices.

The Monthly Wages and Salaries Survey is an employer-based survey. Its average weekly earnings of regular pay and total pay (including bonuses) in the construction industry shows a growth of 50% and 48%, respectively, between 2000 and 2012. This suggests bonuses shrank slightly compared to regular pay.

One reason that may explain the difference between the ONS indices and the national wage agreement–based indices is a drop in working hours of construction workers over time. This appears to be part of the explanation. The two hourly indices collected from the Labour Force Survey and Annual

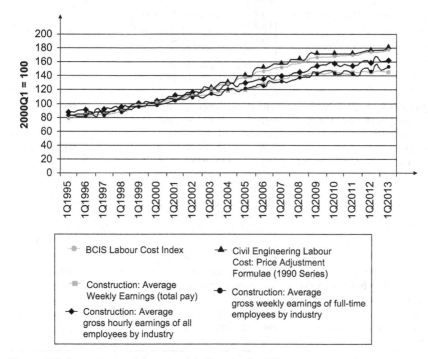

Figure 5.5 Construction labour cost indices in the UK

Survey of Hours and Earnings showed a higher growth (56% and 55%) compared to the weekly earnings indices (49% and 45%) between 2000 and 2012. This is consistent with the 5.2% drop in average weekly hours of work in the construction industry collected by the Labour Force Survey. The experimental index of labour costs per hour published by ONS also reports a 55% increase in labour costs per hour in construction between 2000 and 2012.

Another possible reason is the change in composition of the construction labour force. If the proportion of low-skill construction workers increases over time, the average earnings growth would be lower than the rate of increase in the hourly rates in the national labour agreement. This would require a significant change in the composition to explain the difference, and if such composition change occurs, one would then question the fixed

Table 5.4 Construction labour cost indices in the UK

Labour Cost Indices	Source	Growth rate between 2000Q1 and 2012Q1
BCIS Labour Cost Index	Price Adjustment Formulae (Series 2); all Wage Agreement Bodies listed in table 2	72%
Civil Engineering Labour Index	Price Adjustment Formulae (1990 Series); Civil Engineering Construction Conciliation Board for Great Britain (now Construction Industry Joint Council)	75%
Construction Average Weekly Earnings (Total Pay)	Monthly Wages & Salaries Survey [employer-based survey]	48%
Construction Average Weekly Earnings (Regular Pay)	Monthly Wages & Salaries Survey [employer-based survey]	50%
Median Gross Weekly Earnings	Annual Survey of Hours and Earnings [employer-based survey]	45% [April 2000 to April 2012]
Median Hourly Earnings excluding Overtime	Annual Survey of Hours and Earnings [employer-based survey]	55% [April 2000 to April 2012]
Construction Average Gross Weekly Earnings of Full-Time Employees	Labour Force Survey [household-based survey]	49%
Construction Average Gross Hourly Earnings of all Employees	Labour Force Survey [household-based survey]	56%

Table 5.5 Average Hourly Pay (excluding overtime) as reported in Annual Survey of Hours and Earnings

	Generic	Trade Specific			
	Skilled Construction & Building Trades	*Bricklayers and masons*	*Roofers, roof tilers and slaters*	*Carpenters and joiners*	*Painters and decorators*
2000	£7.66	£7.77	£7.09	£7.71	£7.48
2012	£12.01	£11.49	£10.96	£11.05	£10.53
Growth	57%	48%	55%	43%	41%

weighting in the construction cost index, which would overstate the labour cost inflation by not allowing for substitution. However, Franklin and Mistry's (2013) data suggests that construction labour quality has marginally improved (by around 2%) between 2000 and 2012.

There is no good measure of the labour cost holding the quality and composition of the labour force in construction constant, but a comparison of the hourly rate of a few occupations in the construction industry between the 2000 and 2012 Annual Survey of Hours and Earnings provides an intriguing result, shown in Table 5.5.

Generally the growth of the hourly pay of the generic 'skilled construction and building trades' is in line with the ONS hourly earnings statistics (Median Hourly Earning excluding Overtime and Construction Average Gross Hourly Earnings of all Employees in Table 5.4), while the hourly rates for specific trades displayed a slower growth. This seems to suggest that the wages of the traditional trades covered by the national wage agreements (as reflected in BCIS Labour Cost Index and Civil Engineering Labour Index in Table 5.4) would grow more slowly, not faster, than the ONS hourly earnings statistics.

Recommendations for construction cost indices

With regard to the CCI, a detailed study of the labour cost indices is recommended. ONS's household- and employer-based surveys report a lower growth than the national wage agreement–based indices. Focus should be given to analysing changes in the composition and skill levels of the labour force.

The Price Adjustment Formula–based weightings were last updated in 1995 and could benefit from updating.

Concluding remarks

In this chapter, the compilation methods used to produce the three best-known tender price indices for new buildings in Britain and the two sets of construction cost indices in Britain were surveyed. Having reviewed the compilation methods and the sources of data, we consider that the TPIs published in Britain tend to overstate the inflation of contract prices. The reason is that TPIs only measure the inflation of the traditional trade items such as those relating to the structure and the internal finishes works under the conventional BQ procurement route, but M&E services items and proprietary items such as curtain walls, which are subject to higher productivity growth, are not measured in the indices. Moreover, quality of building work such as energy efficiency and safety driven by building regulations tends to improve over time, and the lack of measurement of quality will tend to overstate the prices over time. In theory, the current expenditure-weighted nature of TPI will tend to understate inflation, but the effect will be limited by new items not being matched to items in the dated schedule of rates.

Measuring the price movement of M&E items and broadening the sample base to design and build contracts are two areas well worth pursuing to restore the representative nature of the indices. For design and build, acquiring access to contract price information such as contract cost plans and the effectiveness of using such cost information to produce TPI deserve further study. Because of the diversity of the M&E items used to achieve comparable performance, it is difficult to stretch the current matched-item index method to measure the price movement of the M&E items. Therefore, there is a need to depart from the presently adopted method, and a hedonic index is an appealing alternative. Although the indices may become less consistent than the existing pure item-matching method, this is a trade-off for improving representativeness.

The CCIs, with an infrequent revision of the base basket, suffer the general base-basket-index shortcoming of overstating inflation. The CCIs are also not designed to reflect productivity growth. Looking at the components, the reliability of the labour cost components must be questioned, and an in-depth study with specific focus on the change in the composition and skill levels of the construction labour force is recommended. These three factors – base-basket weights, no reflection of productivity growth and the labour cost components of the CCIs – all tend to give the indices an upward bias.

References and Further Reading

Akintoye, A. and Skitmore, M. (1994) A Comparative Analysis of Three Macro Price Forecasting Models. *Construction Management and Economics*, 12 (3), 257–270.

Akintoye, A., Bowen, P. and Hardcastle, C. (1998) Macro-economic Leading Indicators of Construction Contract Prices. *Construction Management and Economics*, 16 (2), 159–175.

Allen, S. (1985) Why Construction Industry Productivity is Declining. *Review of Economics and Statistics*, 67 (4), 661–669.

Allen, S. (1989) Why Construction Industry Productivity is Declining: Reply. *Review of Economics and Statistics*, **71** (3), 547–549.

Azzaro, D. W. (1976) *Cost Study F5: Measuring the Level of Tender Prices*. (London: Building Cost Information Service).

Ball, A. and Allen, A. (2003) The Introduction of Hedonic Regression Techniques for the Quality Adjustment of Computing Equipment in the Producer Prices Index (PPP) and Harmonised Index of Consumer Prices (HICP). *Economic Trends*, (592), 30–36.

BCIS (1983) *Cost Study F34: The Tender Price Index-an Update of BCIS Methodology*. (London: Building Cost Information Service).

BCIS (1997) *BCIS Building Cost Index Models*. (Surrey: Building Cost Information Service).

Berry, S., Levinsohn, J. and Pakes, A. (1995) Automobile Prices in Market Equilibrium. *Econometrica*, **63** (4), 841–890.

BIS (2012) *Construction Resources Cost Indices Notes and Definitions*. Department for Business, Innovation & Skills. Available at: www.gov.uk/government/publications/construction-resource-cost-indices-notes-and-definitions: Accessed July 2013.

BIS (2013) *Results of BIS Construction on the Uses of Construction Price and Cost Indices*. (London: Department for Business Innovation and Skills).

Blackman, I. and Picken, D. (2010) Height and Construction Costs of Residential High-Rise Buildings in Shanghai. *Journal of Construction Engineering and Management*, **136** (11), 1169–1180.

Bowley, M. and Corlett W. (1970) *Report on the Study of Trends in Building Prices*. (London: Ministry of Public Building and Works, Directorate of Research and Information).

Bresnahan, Timothy F. and Gordon, Robert J. (eds.) (1996) *The Economics of New Goods*. (Chicago: University of Chicago Press).

Chan, S. and Park, M. (2005) Project Cost Estimation Using Principal Component Regression. *Construction Management and Economics*, **23** (3), 295–304.

Chen, W. and Huang, Y. (2006) Approximately Predicting the Cost and Duration of School Reconstruction Projects in Taiwan. *Construction Management and Economics*, **24** (12), 1231–1239.

Cole, R., Chen, Y. C., Barquin-Stolleman, J., Dulberger, E., Helvacian, N. and Hodge, J. (1986) Quality-Adjusted Price Indexes for Computer Processors and Selected Peripheral Equipment. *Survey of Current Business*, **66** (January), 41–50.

Corrado, Carol and Slifman, Lawrence (1999) Decomposition of Productivity and Unit Costs. *American Economic Review*, **89** (2), 328–332.

Davis Langdon Management Consulting (2010) *Review of BERR/BIS Construction Price and Cost Indices*. (London: Davis Langdon).

Department of the Environment (1995) *Price Adjustment Formulae for Construction Contracts*. (London: HMSO).

Dyer, Bryan, Goodrum, Paul M. and Viele, Kert (2012) Effects of Omitted Variable Bias on Construction Real Output and Its Implications on Productivity Trends in the United States. *Journal of Construction Engineering and Management*, **138** (4), 558–566.

Elhag, T. and Boussabaine, H. (1999) Tender Price Estimation: Neural Networks vs Regression Analysis. In: *Proceedings of the RICS Construction and Building Research Conference*. University of Salford, 114–123.

Emsley, M., Lowe, D., Duff, A., Harding, A. and Hickson, A. (2002) Data Modelling and the Application of a Neural Network Approach to the Prediction of Total Construction Costs. *Construction Management and Economics*, 20 (6), 465–472.

EUROSTAT (1996) *Methodological Aspects of Construction Price Indices*. (Luxembourg: Office of Official Publications of the European Communities).

Fellows, R. (1991) Escalation Management: Forecasting the Effects of Inflation on Building Projects. *Construction Management and Economics*, 9 (2), 187–204.

Fenwick, D. and Wingfield, D. (2005) Methodological Improvements to the Retail Prices Index and Consumer Prices Index from February 2005. *Economic Trends*, (616), 74–80.

Fisher, I. (1921) The Best Form of Index Number. *Quarterly Publication of the American Statistical Association*, 17 (133), 533–537.

Fleming, M. C. (1965) Costs and Prices in the Northern Ireland Construction Industry. *Journal of Industrial Economics*, 14 (1), 42–54.

Fleming, M. C. (1966) The Long-Term Measurement of Construction Costs in the United Kingdom. *Journal of the Royal Statistical Society, Series A (General)*, 129 (4), 534–556.

Fleming, M. C. and Tysoe, B. A. (1991) *Spon's Construction Cost and Price Indices Handbook*. (London: E and FN Spon).

Franklin, Mark and Mistry, Priya (2013) *Quality-Adjusted Labour Input: Estimates to 2011 and First Estimates to 2012*. Office for National Statistics. Available at: www.ons.gov.uk/ons/dcp171766_317119.pdf.

Goodrum, P. M., Hass, C. T. and Glover, R. W. (2002) The Divergence in Aggregate and Activity Estimates of US Construction Productivity. *Construction Management and Economics*, 20 (5), 415–423.

Gordon, R. (1968) A New View of Real Investment in Structure. *Review of Economics and Statistics*, 50 (4), 417–428.

Gordon, Robert J. (1971) Measurement Bias in Price Indexes for Capital Goods. *Review of Income and Wealth*, 17 (June), 121–174.

Gordon, Robert J. (1979) *Energy Efficiency, User Cost Change and the Measurement of Durable Goods Price*. NBER Working Paper No. 408. (Cambridge, Mass.: National Bureau of Economic Research).

Gordon, Robert J. (2005) *Apparel Price 1914–93 and the Hulten/Brueghel Paradox*. NBER Working Paper No. 11548. (Cambridge, Mass.: National Bureau of Economic Research).

Griliches, Z. (1994) Productivity, R&D and the Data Constraint. *American Economic Review*, 84 (1), 1–23.

Gruneberg, S. (2007) Performance-Based Contracting: An Alternative Approach to Transacting in Construction. *Construction Management and Economics*, 25 (2), 111–112.

Gullickson, William and Harper, Michael J. (1999) Possible Measurement Bias in Aggregate Productivity Growth. *Monthly Labor Review*, 122 (2), 47–67.

Guo, K., Minchin, E. and Ferragut, T. (2005) The Shift to Warranties and Performance Specifications: What of Method Specifications? *Construction Management and Economics*, 23 (11), 953–963.

Haskell, Preston H. (2004) *Construction Industry Productivity: Its History and Future Direction*. White Paper by Haskell, America's Design-Build Leader.

Hausman, Jerry (1999) Cellular Telephone, New Products and the CPI. *Journal of Business & Economic Statistics*, **17** (2), 188–194.

Hwang, Seokyon (2009) Dynamic Regression Models for Prediction of Construction Costs. *Journal of Construction Engineering and Management*, **135** (5), 360–367.

Hwang, Seokyon (2011) Time Series Models for Forecasting Construction Costs Using Time Series Indexes. *Journal of Construction Engineering and Management*, **137** (9), 656–662.

Ive, G., Gruneberg, S., Meikle, J. and Crosthwaite, D. (2004) *Measuring the Competitiveness of the UK Construction Industry: Volume 1.* (London: Department of Trade and Industry).

Ji, S., Park, M. and Lee, H. (2010) Data Preprocessing-Based Parametric Cost Model for Building Projects: Case Studies of Korean Construction Projects. *Journal of Construction Engineering and Management*, **136** (8), 844–853.

Kim, Gwang-Hee, An, Sung-Hoon, and Kang, Kyung-In (2004) Comparison of Construction Cost Estimating Models Based on Regression Analysis, Neural Networks, and Case-Based Reasoning. *Building and Environment*, **39** (10), 1235–1242.

Lowe, D., Emsley, M. and Harding, A. (2006a) Predicting Construction Cost Using Multiple Regression Techniques. *Journal of Construction Engineering and Management*, **132** (7), 750–758.

Lowe, D., Emsley, M. and Harding, A. (2006b) Relationships between Total Construction Cost and Project Strategic, Site Related and Building Definition Variables. *Journal of Financial Management of Property and Construction*, **11** (3), 165–180.

Lowe, D., Emsley, M. and Harding, A. (2007) Relationships between Total Construction Cost and Design Related Variables. *Journal of Financial Management of Property and Construction*, **12** (1), 11–23.

Marshall, A. (1887) Remedies for Fluctuations of General Prices. *Contemporary Review*, **51**, 355–375. Reprinted in Pigou, A. C. (ed.) *Memorials of Alfred Marshall*, 188–211 (London: Macmillan).

McCaffer, R., McCaffrey, M. J. and Thorpe, A. (1983) The Disparity between Construction Cost and Tender Price Movements. *Construction Papers*, **2** (2), 17–27.

Meikle, J. (2001) A Review of Recent Trends in House Construction and Land Prices in Great Britain. *Construction Management and Economics*, **19** (3), 259–265.

Mitchell, R. (1971) A Tender-Based Building Price Index. *Chartered Surveyor*, **104** (1), 34–36.

Musgrave, J. (1969) The Measurement of Price Changes in Construction. *Journal of the American Statistical Association*, **64** (327), 771–786.

National Consultative Council Working Group on Indices for Building Contracts (1979) *Price Adjustment Formulae for Building Contracts.* (London: Her Majesty's Stationery Office).

Ng, T., Cheung, S., Skitmore, M. and Wong, T. (2004) An Integrated Regression Analysis and Time Series Model for Construction Tender Price Index Forecasting. *Construction Management and Economics*, **22** (5), 483–493.

Nordhaus, W. (1997) Do Real-Output and Real-Wage Measures Capture Reality? The History of Lighting Suggests Not. In: Bresnahan, T. F. and Gordon, R. J. (eds.) *The Economics of New Good*, 29–66. (Chicago: University of Chicago Press).

OECD and EUROSTAT (2001) *Main Economic Indicators – Sources and Methods: Construction Price Indices.* (Paris: Statistics Directorate, Organisation for Economic Co-operation and Development).

Ohta, M. (1975) Production Technologies of the U.S. Boiler and Turbogenerator Industries and Hedonic Price Indexes for their Products: A Cost-Function Approach. *Journal of Political Economy*, 83 (1), 1–26.

Pakes, A. (2003) A Reconsideration of Hedonic Price Indexes with an Application to PCs. *American Economic Review*, 93 (5), 1578–1596.

Pieper, P. (1989) Why Construction Industry Productivity Is Declining: Comment. *Review of Economics and Statistics*, 71 (3), 543–546.

Pieper, P. (1991) Measurement of Construction Prices: Retrospect and Prospect. In: Berndt, E. and Triplet, J. (eds.) *Fifty Years of Economic Measurement: The Jubilee of the Conference on Research in Income and Wealth*, 239–272. (Chicago: University of Chicago Press).

PSA QSSD (late 1980s) *Tender Level Yardstick Instructions.* Property Service Agency Quantity Surveyors Services Division, London. Unpublished internal document.

Richardson, C. and Dolling, M. (2005) Imputed Rents in the National Accounts. *Economic Trends*, 617, 36–41.

RICS (2006) *Contracts in Use: A Survey of Building Contracts in Use during 2004.* (London: Quantity Surveying and Construction Faculty, Royal Institution of Chartered Surveyors).

RICS (2012) *Contracts in Use: A Survey of Building Contracts in Use during 2010.* (London: Built Environment, Royal Institution of Chartered Surveyors).

Silver, M. and Heravi, S. (2004) Hedonic Price Indexes and the Matched Models Approach. *Manchester School*, 72 (1), 24–49.

Somerville, C. T. (1999) Residential Construction Costs and the Supply of New Housing: Endogeneity and Bias in Construction Cost Indexes. *Journal of Real Estate Finance and Economics*, 18 (1), 43–62.

The Steering Group on Price Fluctuations Formulae (1969) *Formulae Methods of Price Adjustment on Building Contracts: Report on Price Fluctuations Formulae for the Building Industry.* (London: National Economic Development Office).

Stoy, C. and Schalcher, H. (2007) Residential Building Projects: Building Cost Indicators and Drivers. *Journal of Construction Engineering and Management*, 133 (2), 139–145.

Thalmann, Philippe (1998) A Low-Cost Construction Price Index Based on Building Functions. *Proceedings of the 15th International Cost Engineering Congress.* (Rotterdam, the Netherlands: ICEC).

Triplett, J. (2004) *Handbook on Hedonic Indexes and Quality Adjustments in Price Indexes: Special Application to Information Technology Product.* OECD Science Technology and Industry Working Papers, 2004/9. (Paris: OECD Publishing).

UK Statistics Authority (2011a) *Assessment of Compliance with the Code of Practice for Official Statistics: Construction Price and Cost Indices produced by the Department for Business, Innovation and Skills.* Assessment Report 95. (London: UK Statistics Authority).

UK Statistics Authority (2011b) *Assessment of Compliance with the Code of Practice for Official Statistics: Statistics on Building Materials and Components produced by the Department for Business, Innovation and Skills.* Assessment Report 168. (London: UK Statistics Authority).

Wolfson, Michael C. (1999) New Goods and the Measurement of Real Economic Growth. *Canadian Journal of Economics*, 32 (2), 447–470.

Yu, Marco K. W. and Ive, Graham (2008) The Compilation Methods of Building Price Indices in Britain: A Critical Review. *Construction Management and Economics*, 26 (7), 693–705.

Chapter 5 Appendix

This section describes the three well-known methods of constructing price indices and their characteristics. These methods can produce bilateral price index numbers for two periods. Multilateral price index numbers for many periods can be derived from these bilateral indices. We will return to the methods of constructing multilateral price indices from bilateral indices later.

These three types of price indices are all the ratios of the weighted average of the prices in the reference period to the weighted average of the prices in the base period. The different ways to 'weight' the prices set them apart from each other.

Laspeyres price index

Laspeyres price index is a base weight index. The relative quantities of the base period provide the weighting for the respective prices. The following is the formula for calculating the Laspeyres Price Index:

$$\frac{\sum_j p_{tj} \times q_{oj}}{\sum_j p_{oj} \times q_{oj}}$$

where p_{tj} is the price of the jth good at time t (reference period); p_{oj} is the price of jth good at time 0 (base period); q_{oj} is the quantity of the jth good at time 0 (base period).

For example, we construct the price index of 'cereals' comprising rice and wheat. In the base year, the economy produces 1,000kg of wheat and the price is £1/kg. It also produces 500kg of rice at £2/kg. In the reference year, the economy produces 1,000kg of wheat and the price is £3/kg. It also produces 1,000kg of rice at £3/kg. Setting the index at the base year as 100, the Laspeyres price index of cereals in the reference year is

$$\frac{£3 \times 1000 + £3 \times 500}{£1 \times 1000 + £2 \times 500} \times 100 = 225$$

In the ideal case, the goods found in the base period are matched with the exact goods found in the reference period. Therefore, a Laspeyres index has a good control of the quality of the goods being indexed. However, it does not take into account the quantities in the reference period, and people will tend to substitute a cheaper good for a more expensive one in case of a relative price change. In our example, as the price of rice has fallen relative to the price of wheat (from a rate of exchange of 1kg of rice for 2kg of wheat to a rate of 1kg of rice to 1kg of wheat), so consumers have switched towards consuming relatively more rice (from half as much rice as wheat to equal quantities). As a result, the Laspeyres index is often criticized as subject to substitution bias (failure to capture substitution effects) which overstates the inflation.

Paasche price index

A Paasche price index is a current weight index, and its generic formula is as follows:

$$\frac{\sum_j p_{tj} \times q_{tj}}{\sum_j p_{oj} \times q_{tj}}$$

The only new notation is q_{tj}, which stands for the quantity of the jth good at time t (reference period).

This is more suitable for deflating output than the Laspeyres index, as the current outputs are used as the weightings. However, it is criticized as understating the inflation as it does not reflect the choice of goods under the base period prices.

The Paasche price index of our rice and wheat example is as follows:

$$\frac{£3 \times 1000 + £3 \times 1000}{£1 \times 1000 + £2 \times 1000} \times 100 = 200$$

Fisher ideal index

If one index tends to overstate inflation and the other tends to understate inflation, it is natural to take the average of them as a better approximation of the true measure of inflation (Fisher 1921). Irving Fisher exactly suggested this and dubbed it the "best form of index number". The formula for it is as follows:

$$\sqrt{\frac{\sum_{j} p_{tj} \times q_{oj}}{\sum_{j} p_{oj} \times q_{oj}} \times \frac{\sum_{j} p_{tj} \times q_{tj}}{\sum_{j} p_{oj} \times q_{tj}}}$$

It, in theory, should be a better measure of the true inflation. However, the advantage comes with a cost, which requires the information of quantities at both base and reference periods.

The Fisher ideal index of our rice and wheat example is as follows:

$$\sqrt{\frac{£3 \times 1000 + £3 \times 500}{£1 \times 1000 + £2 \times 500} \times 100 \times \frac{£3 \times 1000 + £3 \times 1000}{£1 \times 1000 + £2 \times 1000} \times 100} = 212$$

All these three methods assume complete price and quantity data for all goods is available. However, in reality, new goods enter the market and old goods drop out. Certainly it is less than straightforward to ascertain how much a laptop computer should be priced at in 1900 as well as how much a Ford Model T should be in 2007.

Editorial comment

Amongst the problems that confront those who carry on any sort of research work, particularly in the construction industry, is that of obtaining reliable and consistent data. Even within a single country, it is often difficult to get data, and even when data is available, it is often compiled and presented in different ways. Indeed, the industry and its components are often measured in quite different ways by different people, and this adds a layer of complexity to any attempt to compare data from different sources.

There have been attempts to standardize some aspects of the measurement of construction, and we are familiar with documents such as the various standard methods of measurement of buildings and building work, ranging from those that prescribe how building areas will be measured and how elemental costs will be measured and recorded to those that provide great detail regarding the measurement of work items in bills of quantities. These rules provide a common basis for the measurement and presentation of specific types of building data, but they are generally national publications that have little relevance outside their country of origin. Even within a country, there is a lot of measurement of construction undertaken that follows no set rules. Estimators are often responsible for measurement for tendering purposes, and while there may be some standardization of approach within a firm, that will seldom be exactly the same as the approach taken in other places.

At a national level, there are many ways to measure construction activity or output with little consistency in the way that, for example, work to existing buildings is differentiated from work on new buildings. Renovation and maintenance of existing buildings generally represents a substantial part of total construction, but the proportion will vary from country to country, and such work may or may not be included in overall measures of construction activity. Such differences in what is measured and what is not, what is separated and what is included, and so on, make comparisons difficult, as it will often be necessary to try to adjust data from different sources so that comparisons of like with like are achieved.

Collecting and compiling accurate and reliable national statistics is an expensive exercise that is beyond the capacity of many small and/or developing countries; as a result it is often difficult, if not impossible, to get anything more than sketchy data for even fundamental items such as total construction output or number of people employed in construction. Researchers are generally looking for finer detail such as output by sector (e.g. residential v. nonresidential construction), yet that sort of detail is often not captured, and even if it is, it may be categorized differently in different countries. Informal construction and/or the 'black' economy where building work is paid for in cash or kind and thus goes unrecorded adds extra dimensions of uncertainty.

In spite of these difficulties, there are ongoing attempts to use the data that is available to explore many aspects of how the construction industry operates both within countries and across national borders. In the following chapter, the authors look at a number of key issues related to the nature, availability and interpretation of national construction data.

6 Measuring and comparing construction activity internationally

Jim Meikle and Stephen Gruneberg

Introduction

The notion that there is a predictable relationship between construction and economic development was first put forward by Turin (1969) and subsequently developed by Drewer (1980). Since then, a body of construction industry development theory has been built on that literature, including Bon (1992, 2000), Crosthwaite (2000), Ofori and Han (2003), Ruddock and Lopes (2006) and Gruneberg (2009).

One would expect wealthier economies to have larger construction outputs. Turin's suggestion, however, was that, as an economy grows, the proportion of construction in national output would increase until, at some point, it levels out. Bon (1992), using the same kind of data, but updated, suggested what has come to be known in the literature as the 'Bon Curve': that, as national output continues on a growth path, after a period of levelling out, construction will tend to decline as a proportion of national output.

Two interlinked phenomena were suggested by Turin and Bon. Firstly, there is a direct relationship between construction value added per capita and GDP per capita and, second, as countries grow wealthier, construction value added tends to increase at a faster rate than growth in the overall economy. Bon's refinement is that, at a certain point, growth in construction value added tends to slow and, eventually, begins to decline as a proportion of GDP. More recently, Gruneberg (2009) has suggested that, while there may be an increase in construction activity in the initial stages of economic development, this may subsequently slow down. Thereafter, construction may rise and fall depending on changing circumstances, but with no particular or predictable pattern.

The studies mentioned and other analyses of construction's contribution to the economy are in terms of construction value added, presumably because data in that form is more or less readily available from national accounts. Moreover, construction data and GDP are invariably converted to a common currency – usually $US – using commercial exchange rates. The use of value-added and commercial exchange rates is questioned in this chapter. The chapter casts a fresh eye on the contribution of construction to national economies and how that contribution varies across countries and over time.

Users and uses

Construction activity is distinct from but obviously related to the property sector. However, it does not include the value of land. Construction activity is also separate from but increasingly overlaps with construction professional services. The cost of at least some construction professional services is often included in construction contracts and, therefore, construction output data, although it may be provided by consultancy firms subcontracted to construction contractors.

The main potential users of reliable national and international construction data include construction companies (contractors and consultants), construction industry clients, property agents and developers, government agencies, researchers and the press. Their needs are varied but largely comprise market or statistical data and analyses, including data on output and new orders, costs and prices and productivity. Their interest in construction data extends to levels (absolute values) and trends in:

- the size of markets for different types of construction work, for example, housing, office buildings and power projects. There is often interest locally and internationally for particularly large programmes of work or for significant falls in the demand for construction work. A number of specialist market intelligence organizations produce construction market reports, usually based on official data.
- prices for the main construction resource inputs. Rapidly rising or falling resource prices are of interest to people inside and outside the industry. Price changes are usually an indication of local demand pressures but sometimes result from international demand for particular resources. Structural steel is an example in recent years. The construction press publishes regular articles on construction price trends, for example, Gardiner and Theobald (2012), Rider Levett Bucknall (2013) and Turner and Townsend (2013); there are also annual – or more frequent – price books in a number of countries, such as Rawlinsons' *Australian Construction Handbook* (2013).
- output prices for particular building types or types of work. Steep increases or decreases in these also usually follow market demand. Data on output prices is often published with data on resource prices. The effects of construction price trends should not be confused with property price trends. In the UK, for example, the effect of land prices on house prices is frequently ignored, and the cause of property price increases is wrongly attributed to the construction industry.
- construction price level differences across countries. This information is provided on a regular basis by international firms of cost consultants, such as Compass International (2013). Usually these include the prices of resources and different types of construction work. Priced items relate to representative local materials and products and projects

that are representative of local practice and standards; and prices are usually brought to a common basis using commercial exchange rates. There have, however, been efforts recently to take a more considered approach, for example, by Faithful and Gould (2013) and Turner and Townsend (2013).

- indicators of national competitiveness and productivity. These are of great interest to industry bodies, national governments and the press and often become a source of national pride or national concern. In fact, accurate, meaningful measures can be elusive, but that does not prevent regular publication of reports providing 'data' ranking countries' performance.

However, because of the ways in which construction is measured and presented, what is measured and how it is presented provide opportunities to use the wrong data or to use it in misleading ways. These issues are discussed later. In addition, it is important that some kind of commentary or critique accompanies published construction data; this is particularly appropriate for the press and the wider dissemination of the data. The opportunities for misunderstanding or misinterpretation – deliberate or not – of construction data are great. More effort needs to be made to explain terms and concepts and how they should be used and interpreted. There also needs to be explicit acknowledgement of the shortcomings in the quality of much construction data.

Measures of construction activity

Two aspects of measuring construction activity are what is measured and how it is presented. A major source of inconsistency in international comparisons is the variety of definitions used, what is included and what is excluded from national measures of construction activity. As has been noted by Meikle and Grilli (1999), construction activity in the national accounts is not consistent either because some activities are included elsewhere in the national accounts (for example, construction design and management services are usually included in professional services), because they are not included at all or because they are only partially included.

A recent unpublished pilot survey undertaken by Meikle in 2011 on behalf of the African Development Bank (AfDB) revealed an even more complicated situation in African countries. Unlike most developed countries, the majority of countries in Africa have a high proportion of construction activity that is informal or undertaken by households that is often unrecorded. Recorded construction activity is usually broken down into work by registered and unregistered contractors and households and may or may not include work by small firms or individual tradesmen and may or may not include very small projects. Surveys, where they are undertaken, are usually sample surveys, and the value of most work is estimated based on material consumption or other indicators. The 2011 AfDB survey indicated that different things are measured in different countries and in different ways. In some countries, only work by registered contractors is included in the national accounts; in others, virtually everything that could be included is.

There are three commonly used definitions of construction activity: construction value added, gross construction output and contractors' output. Construction value added and gross construction output are national accounts concepts and are supposed to include all activity that produces construction output, including work by contractors, work by government agencies, work by households and informal (black- or grey-economy) construction output. The third measure, contractors' output, is the gross value of construction work produced by construction contractors. Because the output of construction is so heterogeneous, aggregate quantity measures of construction are not feasible, and using monetary values is the only option.

Construction value added is the component of GDP that measures the value added by construction firms to inputs from other parts of the economy to produce gross construction output. Typically, it includes compensation of employees, depreciation of capital investments and gross operating surplus, crudely, labour, depreciation of capital and profit. The concept of value added avoids double counting in national accounts, as the aggregate of all value added by all economic sectors is the measure of GDP.

Gross construction output, the second measure, appears as a component of national investment, Gross Fixed Capital Formation (GFCF). It is the aggregate of construction value added and the value of contributions from all other parts of the economy that produce construction works. Gross output is the sector's turnover or the amount paid by its customers for its products. Strictly speaking, investment in construction excludes construction repair and maintenance activity, although it includes major refurbishments and extensions, but this distinction is not always observed. The 2005 International Comparison Program (ICP) (International Bank for Reconstruction and Development 2008), for example, breaks down GDP in expenditure terms, by investment and consumption, but includes all construction expenditure as investment.

The third measure, contractors' output, is the total work undertaken by registered contractors and, in some cases, estimates of work by public-sector enterprises and unregistered contractors. Own-account work by commercial organizations is included in the organizations' output.

Construction in the national accounts

Table 6.1 sets out the breakdown of construction activity in national accounts required by the United Nations International Standard Industrial Classification (ISIC).

Table 6.1 SIC activity categories for construction

F	CONSTRUCTION				
41	Construction of buildings	41.1	Development of building projects	41.10	Development of building projects
		41.2	Construction of residential and nonresidential buildings	41.20	Construction of residential and nonresidential buildings 41.20/1: Construction of commercial buildings 41.20/2: Construction of domestic buildings
42	Civil engineering	42.1	Construction of roads and railways	42.11	Construction of roads and motorways
				42.12	Construction of railways and underground railways
				42.13	Construction of bridges and tunnels
		42.2	Construction of utility projects	42.21	Construction of utility projects for fluids
				42.22	Construction of utility projects for electricity and telecommunications
		42.9	Construction of other civil engineering projects	42.91	Construction of water projects
				42.99	Construction of other civil engineering projects n.e.c.
43	Specialized construction activities	43.1	Demolition and site preparation	43.11	Demolition
				43.12	Site preparation
				43.13	Test drilling and boring
		43.2	Electrical, plumbing and other construction installation activities	43.21	Electrical installation
				43.22	Plumbing, heating and air-conditioning installation
				43.29	Other construction installation
		43.3	Building completion and finishing	43.31	Plastering
				43.32	Joinery installation
				43.33	Floor and wall covering
				43.34	Painting and glazing 43.34/1: Painting 43.34/2: Glazing

(*Continued*)

Table 6.1 (Continued)

F	CONSTRUCTION		
		43.39	Other building completion and finishing
43.9	Other specialized construction activities	43.91	Roofing activities
		43.99	Other specialized construction activities n.e.c. 43.99/1: Scaffold erection 43.99/2: Specialized construction activities (other than scaffold erection) n.e.c.

The national accounts work categories are intended to include all construction activity by all agencies, not only work by contractors and direct works organizations. The categories, however, are not particularly helpful or complete. There are only two subcategories for building work (commercial and residential) when building probably represents more than half and often more than three quarters of construction activity in most countries; there is rather more detail for civil engineering work (seven subcategories); and there is no separation of public and private work or new work and work to existing buildings.

It is fairly common but not universal in national accounts, where there is a breakdown of construction output, to provide data on residential, non-residential and civil engineering work. This is equivalent to ISIC categories 41.20/2, 41.20/1 and 42. It is often not possible, however, to obtain reliable data on construction output by trade (ISIC category 43) or preconstruction activities (ISIC category 41). There is a real need to review and revise the ISIC categories to reflect statistical, industry and other needs and modern practice.

Normalizing construction data

In order to compare value data on construction activity within countries, it is necessary to be able to adjust for differences in purchasing power in terms of type of work as well as the date of construction, location and even qualitative differences. This is done with different levels of reliability in different countries. In the UK, for example, a number of methods are used, such as cost and price indices, to bring construction values for different types of work undertaken at different times to a common price basis.

Spatial variation factors, the focus of this chapter, can be used to adjust for project location. Projects in dense urban or remote locations tend to have higher price levels than projects on open sites adjacent to sources of labour and materials, though other factors can come into play. For example, projects in the South-east of England tend to be more expensive than equivalent projects in the North-east. The purpose of work type and temporal and spatial adjustment factors is to permit comparisons to be made that are based on a common value. As noted in other chapters, the difficulties of making international construction comparisons at a point in time are compounded by the need to use currency convertors based on exchange rates.

Purchasing power parities (PPPs) are an alternative to exchange rates and are applied generally to the whole economy; PPPs are also available for components of the economy. The origin, development, production and use of PPPs are described in detail elsewhere in this book. It is sufficient here to note that PPPs for a range of activities, including construction, are published by the World Bank, currently every five years or so, but the intention is for this to be more frequent in the future. Table 6.2 sets out exchange rates, GDP PPPs and PPPs for selected components of GDP, all based on US$ = 1.00, for a selection of countries from different regions, income levels and population sizes.

In higher-income countries, PPPs are typically higher than exchange rates. In middle- and lower-income countries, exchange rates tend to be higher than PPPs. This means that in rich countries, the size of economies and the volume of national output tend to be overstated and in poorer countries,

Table 6.2 Exchange rates and PPPs for selected countries, 2005

Country	Exchange rate to $US	GDP PPP	Investment PPPs		Consumption PPPs	
			Construction	Machinery and equipment	Food	Clothing
Argentina	2.90	1.27	1.10	3.26	1.70	1.80
Australia	1.31	1.39	1.39	1.61	1.61	1.38
Benin	527.47	220.00	151.51	684.33	495.42	326.76
China	8.19	3.45	1.93	8.79	5.52	6.86
Denmark	5.99	8.52	9.16	7.59	9.61	8.59
Hungary	199.47	128.51	148.62	221.66	161.99	220.41
Indonesia	9,704.70	3,934.26	2,551.52	11,032.96	5,817.59	3,858.04
Kenya	75.55	30.00	28.80	88.58	54.14	34.73
Singapore	1.66	1.08	0.62	1.77	1.78	1.60
South Africa	6.36	3.87	4.08	6.09	5.53	5.56
UK	0.55	0.65	0.77	0.64	0.71	0.67

Source: Based on ICP results

understated. Because construction activity is not internationally traded, construction PPPs tend to exaggerate these tendencies, whereas machinery and equipment, which are internationally traded, tend to have PPPs that are generally similar to or higher than commercial exchange rates. Food and clothing PPPs tend to be closer to GDP PPPs in higher-income countries and closer to exchange rates in lower-income countries. There are, of course, exceptions to these general statements that may be explained by specific national features. In Singapore, a rich country, for example, construction PPPs are lower than might be expected, because there is a reliance on low-cost foreign construction workers.

PPPs demonstrate the Balassa-Samuelson effect. This states that nontraded elements of the economy – like construction – will have low relative prices in the early stages of economic development but that, over time, their price levels will tend to converge with general prices. Using commercial exchange rates for all economic activities, the effect is not necessarily picked up, and this means that the real volume of construction activity in lower-income countries tends to be understated and that the real volume in higher-income countries tends to be overstated. This is an added reason for using PPPs rather than exchange rates for comparing the volume of construction in one country to that in another. However, PPPs can only really be applied to high-level economic concepts. As PPPs are an average of equivalent baskets of goods, they do not apply to particular products or projects. The next section applies the theoretical framework of PPPs to national construction expenditures.

Analysis of the International Comparison Program 2005 data

The 2005 ICP Results (International Bank for Reconstruction and Development 2008) give nominal or commercial exchange-rate adjusted data for GDP and construction expenditure in $US for 128 countries (30 African, 23 Asian, 10 Confederation of Independent States [CIS], 45 European and OECD, 10 Latin American and Caribbean [LAC], and 10 West Asian); they also give GDP and construction expenditure for the same countries using real or PPP adjusted data. Two distinct features of the ICP data are that it provides data on expenditure, not value added, and that it is brought to a common currency basis using both commercial exchange rates and PPPs, both whole-economy PPPs for GDP and construction-specific PPPs for construction expenditure.

The ICP data set allows for an examination of the effect of the use of PPPs on construction volume and on the relationship between GDP and construction expenditure. There are, of course, concerns about the reliability of such value data and PPPs, but the ICP data set is the largest single-source data set available. It is almost certain that the ICP data is more useful than comparisons based on exchange rates.

Table 6.3 sets out GDP, construction expenditure and construction as a proportion of GDP for the 128 countries in their regional groupings using commercial exchange rates and PPPs. The numbers in brackets indicate the numbers of countries for which there is construction data. Coverage is good for Europe/OECD, Asia, the CIS and West Asia but rather less so for Africa and Latin America and the Caribbean.

Table 6.3 illustrates the effect of using PPPs compared to using exchange rates on both construction volume and construction expenditure as a proportion of GDP. Using PPPs, construction expenditure increases markedly in every region except Europe and the OECD, and construction expenditure as a proportion of GDP increases most in Asia and West Asia. There are increases in Africa and Latin America and the Caribbean, but only relatively small ones, as the increases in construction expenditure in the latter are more or less matched by the increases in GDP. The results with PPPs are almost certainly more credible interpretations of the data than those with exchange rates.

The data, of course, represents regional averages, and countries in each region will vary around the mean. Nevertheless, using averages helps to identify broad regional trends. These are in line with expectations, although it is probable that the regions with more complete data – Asia, CIS, EU/OECD and West Asia – will be more reliable than Africa or Latin America and the Caribbean, where there is less complete coverage.

The differences in Asia and West Asia are a result of overvaluation of price levels using commercial exchange rates, particularly for construction. The reasons for this distinction in the two regions are subtly different. Asia comprises mainly middle-income countries with a few high- and low-income countries, a high proportion of locally or regionally manufactured materials and access to low-cost labour. West Asia also comprises high- and middle-income countries with access to low-cost resources and low-cost foreign workers in the case of the high-income countries. The similarities in

Table 6.3 GDP and construction expenditure in 2005 using exchange rates and PPPs

Regional groups	Using commercial exchange rates			Using PPPs		
	GDP $US bn	Construction expenditure $US bn	Construction as a % of GDP	GDP $US bn	Construction expenditure $US bn	Construction as a % of GDP
Africa (30)	711,511	59,122	8.31	1,572,565	141,712	9.01
Asia (23)	4,892,643	935,332	19.12	12,020,712	3,898,645	32.43
CIS (10)	970,013	100,928	10.40	2,269,157	214,822	9.47
EU/OECD (45)	35,409,507	3,980,555	11.24	34,771,461	3,995,997	11.49
LAC (10)	1,601,716	145,830	9.10	3,078,095	361,447	11.74
West Asia (10)	595,710	54,975	9.23	1,000,643	173,840	17.37
All (128)	44,181,100	5,276,742	11.94	54,712,633	8,786,463	16.06

Source: Based on ICP results

both GDP and construction exchanges rates and PPPs in the European and OECD countries are as expected in developed economies.

The reasons for the decrease in construction expenditure as a proportion of GDP in the CIS states are possibly more complex and deserve further study. Although both GDP and construction may be overvalued using commercial exchange rates, this appears to be more so in construction than in GDP. Using the data in Table 6.3, there is a decrease in the proportion of construction to GDP using PPPs (9.47%) compared to exchange rates (10.40%). The difference, however, is less than 1%, which may not be significant bearing in mind the data quality. Overall, using PPPs for 128 countries, GDP increases by around 29%, but construction expenditure increases by 67%, and construction expenditure as a proportion of GDP increases by 30%.

It seems, therefore, that the volume of construction expenditure using exchange rates compared to PPPs tends to be overstated in developed countries and understated in developing countries. Previous research, based on commercial exchange rates, looked at the relationship between construction output (measured as value added) per capita and GDP per capita. Figures 6.1 and 6.2 present the relationship between GDP per capita and construction expenditure per capita using exchange rates and PPPs.

The trend line in Figure 6.1 indicates that, on average, construction expenditure represents around 12% of GDP using commercial exchange rates. Figure 6.2 uses separate GDP and construction PPPs and indicates that the percentage increases to 15%. The figures reflect Turin's view that there is a direct relationship between GDP per capita and construction expenditure per capita; they also reinforce the theory behind the Balassa-Samuelson effect. They do not, however, strongly support Bon's proposition that the proportion of construction expenditure increases in the early stages of development but then levels off or declines in maturity.

The following figures look in more detail at the data by region, by income level and by population. Figures 6.3 to 6.6 show GDP and construction

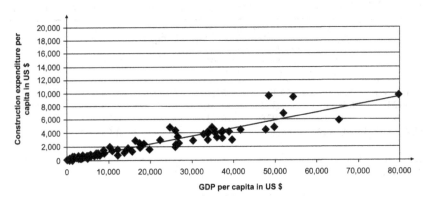

Figure 6.1 GDP and construction expenditure for 128 countries in 2005 in $US using commercial exchange rates

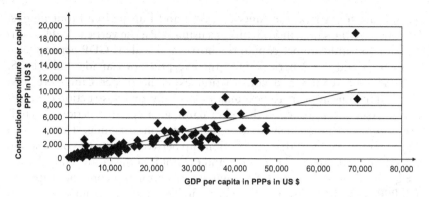

Figure 6.2 GDP and construction expenditure for 128 countries in 2005 in $US using PPPs

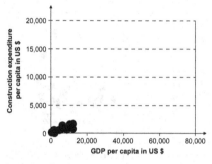

Figure 6.3 GDP per capita and construction expenditure per capita in 2005 $US PPPs by region: Africa

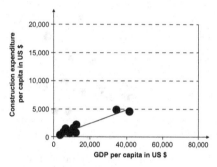

Figure 6.4 GDP per capita and construction expenditure per capita in 2005 $US PPPs by region: the Americas

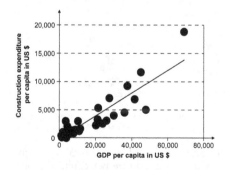

Figure 6.5 GDP per capita and construction expenditure per capita in 2005 $US PPPs by region: Asia

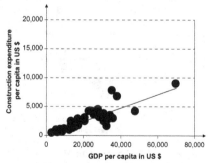

Figure 6.6 GDP per capita and construction expenditure per capita in 2005 $US PPPs by region: Europe

expenditure for Africa, the Americas, Asia and Europe In all four figures, the scales are the same, in $US millions using PPPs; the vertical scale is construction expenditure per capita; the horizontal scale is GDP per capita.

In Africa and the Americas, with the exception of Canada and the US, both GDP per capita and construction expenditure per capita are low. National income appears to be a major constraint on construction expenditure. In Asia and Europe, on the other hand, greater construction expenditure appears to be associated with higher national income. The regional data generally confirms that the relationship between GDP and construction expenditure indicated in Figures 6.1 and 6.2 holds at the more detailed level.

Figures 6.7 to 6.10 group the countries by national income per capita in US$ PPPs. The 128 countries have been divided into five groups, four with 25 countries each and one with 28 (the income bands for these five groups

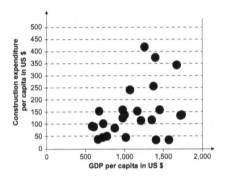

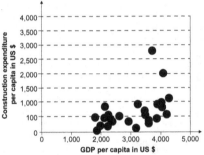

Figure 6.7 GDP per capita and construction expenditure per capita in 2005 $US PPPs by income group: Lowest Income Group

Figure 6.8 GDP per capita and construction expenditure per capita in 2005 $US PPPs by income group: Second-Lowest Income Group

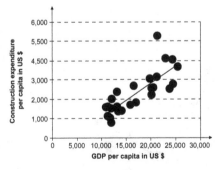

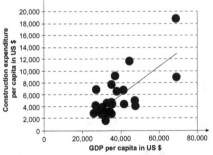

Figure 6.9 GDP per capita and construction expenditure per capita in 2005 $US PPPs by income group: Second-Highest Income Group

Figure 6.10 GDP per capita and construction expenditure per capita in 2005 $US PPPs by income group: Highest Income Group

are set out in Appendix A). In these figures, the middle group has been omitted. The labels for the vertical and horizontal scales are the same as in Figures 6.3 to 6.6, but the values vary to suit the groups.

Generally, as the variance is greater, there is much less sign of a relationship between GDP per capita and construction per capita in the lower income groups (Figures 6.7 and 6.8). Bearing in mind that previous authors have emphasized the importance of national income in predicting construction expenditure, this is surprising. The somewhat better relationship between per capita GDP and construction in the higher income groups (Figures 6.9 and 6.10) may be a result of more geographical homogeneity.

A third grouping of countries was examined – by population size – and a statistical analysis of the results of this exercise and the other two exercises is set out in Table 6.4.

Table 6.4 Statistical tests by region, national income and population size

Construction per head as a function of GDP per head	Number of countries	R^2	Mean	Standard error
By geographic regions				
Africa	30	0.79	72	185.1
The Americas	13	0.93	518	405.9
Asia	42	0.82	361	1,567.2
Europe	41	0.73	1,528	1,009.8
By GDP per capita				
Lowest-income countries	25	0.15	115	99.6
Second lowest	25	0.23	478	543.8
Mid-range	25	0.18	950	424.2
Second highest	25	0.63	2,111	677.9
Highest-income countries	28	0.53	4,556	2,483.7
By population size				
Smallest by population	25	0.80	1,989	2,073.6
Second smallest	25	0.84	949	805.2
Mid-range	25	0.83	611	504.7
Second largest	25	0.93	533	377.8
Largest by population	28	0.75	983	768.9
All countries in survey	128	0.75	440	1,301.6

The table gives the mean and standard errors by geographic region, GDP per capita and population. The regions are shown to be relatively homogenous, as are the groups by population size; while income level in a country is a poor predictor of likely expenditure on construction, regional location and population size are relatively good predictors. In all cases, the standard errors are relatively high compared to the means, leaving wide margins for error in predicting individual country values.

Conclusion

This chapter reviews and comments on some of the main issues concerning the measurement, presentation and comparison of construction activity nationally and internationally. It uses ICP 2005 data to assess the relationships between national income and construction expenditure and concludes that previous theories are, at best, not proven and that more work is needed on defining, measuring and normalizing data on construction activity.

Consistent approaches are required as to what is included in or excluded from construction activity and how variables should be measured and presented. This needs to take account of the data requirements of statisticians, policy makers, international bodies, industry, researchers and others. It is an international issue and needs to be addressed at an international level; construction is too important a sector of the economy to be measured so poorly.

The current ISIC breakdown of construction activity is not particularly helpful to any user groups. It requires distinctions to be made among residential, nonresidential and civil engineering work and between a few subtypes of civil engineering work. It does not distinguish between construction investment (new work and improvements) or construction consumption (repair and maintenance) or among publicly sponsored, privately sponsored and mixed-funded work. Detailed breakdowns of construction activity could also address the different providers of construction output: construction contractors, the informal sector, households and so forth, although it needs to be acknowledged that providers often overlap.

Gross construction output is often a more useful measure of construction activity than construction value added, although the latter has its uses. Gross output data is needed to measure the size and capacity of the sector and its growth rate; value-added measures are needed for measuring construction productivity and the share of on-site activities in gross domestic product. In developed countries, contractors' output can be viewed as broadly representative of construction value; in less developed countries, there is more likely to be greater activity by noncontractors, mainly households.

Construction prices tend to change over time relative to prices in the wider economy both nationally and internationally. Measures of construction activity, therefore, are subject to both quantity and price changes; in these circumstances, construction values are not necessarily good proxies for construction volumes. Indeed, the patterns of national and construction-sector development reported by Turin, Bon and others may be, at least partly, an outcome of relative price trends and may not necessarily reflect real activity. Using the best data available, we cannot confirm Turin and Bon's hypotheses. If regional location and population size are better predictors of construction output per capita than national income, it must be accepted that the basic data as it is currently defined and collected is not as helpful as it might be.

In discussing the use of exchange rates or construction PPPs, a consensus is emerging that exchange rates are poor deflators, or normalizers, in international comparisons of construction activity. Although there are problems with PPPs in general and construction PPPs in particular, there is a rationale for using PPPs in preference to exchange rates for converting construction output and GDP. Furthermore, as PPPs are refined, they can be expected to improve over time, and it is important that their development is encouraged.

Because of the issues and difficulties raised in this chapter, construction activity data – national and international – can be misinformed and misreported. Governmental statistical agencies and industry commentators require improvements in the official definitions of construction and the way data is presented. There is a need for better information on the various measures of construction activity.

In many countries, the construction sector is one of the largest components of the economy. Industrial strategies and policy decisions depend on consistent and reliable measures, and decisions to invest in infrastructure and the built environment remain essential for economic and social development.

References and Further Reading

Best, R. (2012) International comparisons of cost and productivity in construction: a bad example. *Australasian Journal of Construction Economics and Building*, **12** (3), 82–88.

Bon, R. (1992) The future of international construction: secular patterns of growth and decline. *Habitat International*, **16** (3), 119–128.

Bon, R. (2000) *Economic Structure and Maturity* (Aldershot: Ashgate Publishing Ltd).

Compass International (2013) *2013 Global Construction Costs* (Morrisville: Compass International Consultants Inc.).

Crosthwaite, D. (2000) The global construction market: a cross-sectional analysis. *Construction Management and Economics*, **18** (5), 619–627.

Drewer, S. (1980) Construction and development: a new perspective. *Habitat International*, **5** (3), 395–428.

Faithful and Gould (2013) *International Construction Intelligence*, July: www. fgould.com/uk-europe/articles/international-construction-intelligence-ukbase-13/: Accessed: 24 January 2014.

Gardiner and Theobald (2012) *International Construction Cost Survey, December 2012, US$ Version* (London: Gardiner and Theobald). www.gardiner.com/assets/files/files/7df1598c8cfd366d3b879fa7aba67f4bdfe48f46/ICCS%20US$%20Version.pdf: Accessed: 30 December 2013.

Gruneberg, S. (2009) Does the Bon curve apply to infrastructure markets? In: Egbu, C. (ed.) *Proceedings of 26th Annual ARCOM Conference, 6–8 September* (Leeds: Association of Researchers in Construction Management), 33–41.

International Bank for Reconstruction and Development (2008) *Global Purchasing Power Parities and Real Expenditures 2005 International Comparison Program* (Washington: The World Bank).

Meikle, J. L. & Grilli, M. T. (1999) Measuring European construction output: problems and possible solutions. In: *Proceedings of the Second International Conference on Construction Industry Development and First Meeting of CIB Task Group*, 29, 35–44.

Ofori, G. and Han, S. S. (2003) Testing hypotheses on construction and development using data on China's provinces. *Habitat International*, 27 (1), 37–62.

Rawlinsons (2013) *Rawlinsons Australian Construction Handbook*, 31st edition (Rivervale: Rawlinsons Publications).

Rider Levett Bucknall (2013) *International Report, Construction Market intelligence* (London: Rider Levett Bucknall): http://rlb.com/wp-content/uploads/2013/03/rlb-international-report-first-quarter-2013.pdf: Accessed 30 December 2013.

Ruddock, L. and Lopes, J. (2006) The construction sector and economic development: the 'Bon curve'. *Construction Management and Economics*, 24, 717–723.

Turin, D. A. (1969) *The Construction Industry: Its Economic Significance and Its Role in Development*. (London: University College, Environmental Research Group).

Turner and Townsend (2013) *A Brighter Outlook: International Construction Cost Survey 2012* (London: Turner and Townsend): www.turnerandtownsend.com/construction-cost-2012/_16803.html: Accessed 30 December 2013.

Appendix A: Country groupings income ranges by GDP per capita

Income ranges	Number of countries	Income range in $US PPPs
Lowest-income countries	25	590–1,748
Second lowest	25	1,812–4,296
Mid-range	25	4,649–10,692
Second highest	25	11,063–25,527
Highest-income countries	28	26,072–69,270

Population ranges	Number of countries	Population in millions
Smallest by population	25	0.29–3.14
Second smallest	25	3.22–6.81
Mid-range	25	6.85–15.15
Second largest	25	16.28–37.88
Largest by population	28	38.16–1,304

Editorial Comment

Marshall McLuhan introduced the concept of the global village in the early 1960s. His vision of an interconnected world is now very much a reality with virtually instantaneous exchange of information taken for granted in an era of smart phones and high-speed Internet. It is no surprise that the business of procurement of construction projects is now conducted in a global marketplace, with many large construction firms operating not only outside their home country but across many countries.

The construction of major projects has largely evolved into a stratified business, with management separated from onsite activities and much of the work of construction project management organizations now focused on managing people and information. Onsite activity is mostly carried out by local labour under the supervision of a relatively small number of supervisory staff. In an international context, it can be the case that only a few expatriate managers control the work on location, with the parent organization located in another country and the construction site and head office linked by relatively cheap high-speed communications networks.

In this chapter, the authors look at the evolution of international construction and use several approaches to analyse the nature of international contracting. They primarily use data published by *Engineering News-Record* (ENR) to identify who the major players are and where they originally came from. They identify some significant shifts in the mix of contractors now working extensively outside their home countries with a noticeable drop, for example, in the influence of Japanese firms in the top 50 in recent years and a corresponding rise in the prominence of Korean and Chinese contractors.

The authors then use niche theory to examine how the various contractors compete and coexist in the global construction marketplace. They report on an extensive analysis of the international construction context that looks at a variety of aspects of how various firms operate and tailor what they do to meet the demands of particular market segments. The overall result is an interesting investigation of the international construction market and how construction firms exist and operate.

7 Internationalization of the construction industry

Weisheng Lu, Huan Yang and Craig Langston

A quick overview of international construction markets

With the globalization of the world economy, today's construction business is fast becoming an internationally interdependent marketplace. Construction yields on the order of USD4.6 trillion annually and contributed 6.6% to the global gross domestic product (GDP) in 2011 (Davis Langdon 2012). A significant part of that total is attributable to international contractors. With the rise of modern industrialized countries, more and more complex civil engineering projects are being procured, and the increased scale of these projects has provided a launching pad for international construction. For instance, data published by *Engineering News-Record* (ENR) shows that ENR's top 225 international contractors in 2011 earned USD453 billion in revenue from construction projects outside their home countries, which represents a four-fold increase over the USD106.5 billion recorded in 2001 and nearly 20% more than the revenue from the previous year (ENR 2012).

International construction has been defined as the situation in which a company, resident in one country, performs work in another country (Ngowi *et al.* 2005). Although nowadays most big construction companies have both domestic and overseas interests, international construction encourages a global outlook by focusing on competition in overseas markets (Lu *et al.* 2009). Advanced technology, fast transportation, convenient communications, effective knowledge transfer, integrated markets and trade liberalization have all helped to lower traditional barriers and transform construction into a fiercely competitive international marketplace where construction companies rise and fall. A clearer understanding of the competition and performance involved in this market would be helpful to all stakeholders, especially international contractors themselves.

According to ENR, international construction markets can be divided into eight parts: general building, manufacturing, power, water/sewer waste, industrial process/petroleum, transportation, hazardous waste and telecommunications. It can be seen from Figure 7.1 that in the product dimension, transportation and industrial process/petroleum are the most important submarkets. Compared with their revenue just three years earlier, transportation and industrial process/petroleum have shown 16.7% and 17.8% increases, respectively (ENR 2012), and it is these sectors that are the main engine for the recovery of the international construction market. In contrast, the traditional submarket general building has not shown any sign of recovery. ENR (2012) reported USD91.1 billion revenue in the general building market, which is almost 3% lower than the revenue in 2008. Since the barriers to market entry in the general building sector are generally much lower and the competition is much fiercer than in other submarkets, many international contractors are reducing their presence in this sector (ENR 2012).

Geographically, multinational construction can be further divided into six regional segments: North America, Europe, Latin America, Asia, the Middle East and Africa. Similarly, as shown in Figure 7.2, many of these geographical submarkets have shown recovery. ENR (2012) attributed this recovery to pent-up demand due to the stagnant market of the previous three years. In addition, the stimulation policies of the governments of some countries, such as China, may also have boosted their infrastructure markets, thus providing opportunities for international contractors. As the largest potential market that is far from the centre of the global financial crisis, Asia recorded revenue of USD112.2 billion in 2011, a 63.8% increase from 2008 (ENR 2012). In contrast, the European submarket, which was directly hit by the crisis, reported an 11% drop in total international revenue compared to 2008, indicating a slow recovery process.

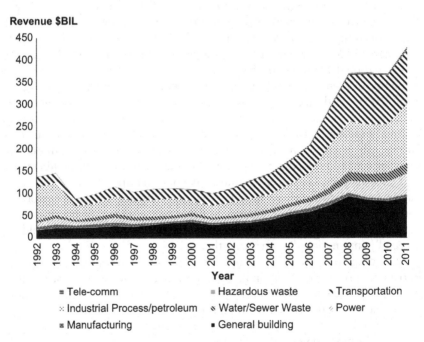

Figure 7.1 Revenue of product markets (USD) 1992–2011 (ENR 2012)

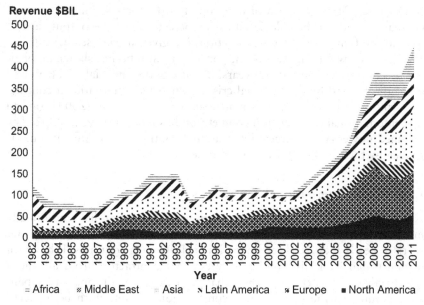

Figure 7.2 Work awarded by regional markets (USD) 1982–2011 (ENR 2012)

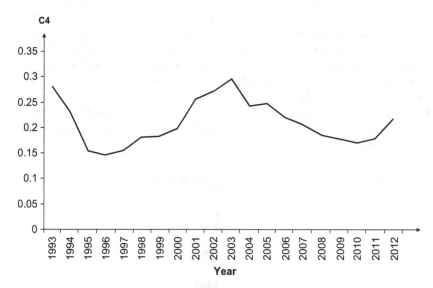

Figure 7.3 Evolution of the concentration ratio 1993–2011 (ENR 2012)

Apart from fluctuations in market revenue, the structure of the market has also been altered by the financial crisis. Many mergers and bankruptcies pertaining to international contractors have been reported by ENR in the past three years. The concentration ratio of the top four international contractors (C4) is an indicator used to reflect the competition in a particular market and can be calculated as the percentage of the outputs attributed to the four largest firms in relation to a given market segment (Clark 1985). As shown in Figure 7.3, the concentration ratio has shown an obvious ascending trend in difficult years. Because of the financial problems and other risks raised by the financial crisis, many small international contractors cannot sustain themselves in a tough market. Ye *et al.* (2009) confirm that international construction competition has intensified since 2002. The regained power of the largest international contractors in difficult times is proven by the trend in concentration ratio.

Evolution of international contractors

Ngowi *et al.* (2005) argue that international construction has traditionally implied companies from advanced induztrialized countries carrying out work in newly induztrialized countries or developing countries. It has been demonstrated by ENR over many years that contractors from advanced induztrialized countries (i.e. those based in America, Europe and Japan) have dominated the international construction market. With better technology, management capacity and financial skills, these contractors were more competitive and able to take advantage of global development. However, more recently, contractors from developing countries have joined the international construction market. This has weakened the dominance of the advanced induztrialized countries. ENR (2012) reported that where the *Engineering News Record* 225 used to be dominated by US, European and Japanese firms, there now are groups of firms on the list based in China, Turkey and the Middle East. It can be seen from Figure 7.4 that international contractors from China and Turkey now occupy more than one third of the positions in the top 225 international contractors, indicating their emergent influence in this market.

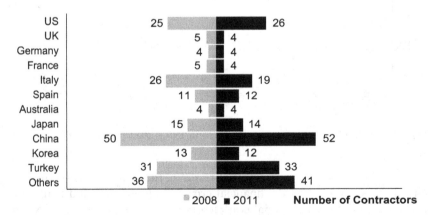

Figure 7.4 Home countries of top 225 international contractors (ENR 2012)

However, compared with the relative stability of the top 225 international contractors, there has been a significant shuffle within the top 50 contractors. These represent the dominant group in the international construction market and are usually contractors from developed countries. Hochtief, a German company, has kept first place in the international market since 2004. The Swedish company Skanska, the French company Vinci and the American company Bechtel have all stayed in the top 10 international contractors for more than 10 years, underlining the dominance of contractors from developed countries. Meanwhile, contractors from developing countries are gradually gaining power in the top group in the international market. As Figure 7.5 shows, the number of Chinese international contractors within the top 50 increased from four in 2008 to nine in 2011. China Communication Construction Group, for example, was ranked 10th in 2011, and its international revenue in that year was USD9,546.9 million, an increase of more than 100% compared with their revenue in 2008 (ENR 2012). Another Asian country, South Korea, has also expanded its position in the top group of international contractors and has contributed to the bloom of the Asian construction market. However, a traditional international participator, Japan, is progressively losing its power in the top group.

Construction companies grow, compete, evolve and perish in the international landscape. Today's international construction is more complicated than ever. The uncertainties of global economic conditions and unstable emerging markets have all cast a cloud over the market. However, risks are always accompanied by opportunities. To gain a competitive position and maintain sustainable growth in this environment, international contractors must achieve a deep understanding of their dynamic environments and shape their strategies according to evolving conditions. The following sections of this chapter will describe a new tool that can be used to

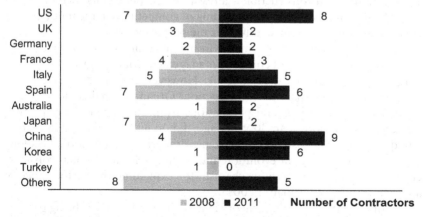

Figure 7.5 Home countries of top 50 international contractors (ENR 2012)

measure and analyse competition and organizational performance in the international construction market.

Niche comparisons: toward a new approach for measuring and analysing competition and organizational performance

Research into various aspects of international construction has been prolific. By and large, these studies can be classified into two categories. In the first category is research in which the trend and framework of the international construction market is analysed from a macro perspective. An example is Bon and Crosthwaite (2001); they investigated future market trends based on their annual worldwide surveys conducted over the course of the last decade. Also, Ye *et al.* (2009) investigated the international construction competition trend over the past 28 years, and Ofori (2003) and Ngowi *et al.* (2005) reviewed the trajectory of the international construction industry as well as the methods for analysing and comparing the performance of international construction contractors. Research in the second category compares companies' strategies in international construction markets across different jurisdictions through analysis from a micro perspective. This has typically been done using a SWOT analysis (strengths, weaknesses, opportunities and threats) based on interviews and case studies (Lu 2010; Lu *et al.* 2009; Zhao and Shen 2008), by using Dunning's eclectic paradigm that emphasizes the ownership, location and internalization advantages of international construction companies (Low and Jiang 2004; Low *et al.* 2004) and by using Porter's competition theories in surveying the competitive advantages of Turkish international construction companies (Öz, 2001). Kale and Arditi (2002) and Korkmaz and Messner (2008) also studied the competitive positions of construction firms mainly based on Porter's competition theories.

Although it is heavily dependent on theories from mainstream management, international construction is a research discipline in its own right that has helped to improve our understanding of competition in the international construction market significantly over the past 20 years.

Moore (1996) saw an economic environment as an ecosystem and claimed that new understandings of company management could be gleaned by studying it from an ecological perspective. However, such studies have rarely been applied to international construction. Organizational ecology, which focuses on organizations and populations from an ecological perspective, was established more than 30 years ago (Baum and Shipilov 2006). Compared with other mainstream management theories, organizational ecology places more emphasis on evolution and natural selection, which considers environment as the primary mechanism for explaining the performance of an organization (Whittington 2001).

The organizational ecology theory is sometimes criticized as being passive, as it places more emphasis on the natural selection process but neglects the initiative of organizations. However, a clearer understanding of the relationship between organizations and the competitive environment in which they operate is necessary, especially in a risky international market. By establishing a proper relationship with the environment, international construction companies are more likely to perform well.

Since international construction is complicated, it is difficult to describe the international status of a company in a holistic sense; hence the introduction of niche theory is warranted. As one of the most important subtheories in organizational ecology, niche theory was initially propagated in the natural bioecology field in relation to multidimensional spaces for organisms or species to survive (Tisdell and Seidl 2004). With its empirical and quantitative characteristics, niche analysis enables a more accurate understanding of the success or failure of an organization by considering its interactions with its environment.

The remainder of this chapter is divided into three sections. First, an NW/O-L (niche width, niche overlap and location) framework is outlined to transfer the conceptual niche to specific constructs. Niche theory is reviewed in conjunction with other theories to identify their similarities and distinctions. Second, the specific parameters included in the NW/O-L framework – niche width (*NW*), niche overlap (*NO*) and location (*L*) – are elaborated upon. Third, the *NW*, *NO* and *L* of top international contractors are calculated. With cluster analysis, they are divided into groups according to their niche. The context and performance of the different groups is then described and compared. The final part of the chapter provides conclusions and suggestions for further research.

Niche theory

The term 'niche' was initially defined in the bioecology field and first introduced into economics as a concept by Hannan and Freeman (1977). Niche in economics is taken as an N-dimensional environment, with each dimension showing the level of relevant environmental condition. Tisdell and Seidl (2004) specified that a niche for a firm is associated with the ability of the firm to stave off competition from other firms and consequently gain a degree of security or comfort. Dimmick *et al.* (2004) suggested that niche theory explains how a company competes and coexists in a limited-resource environment. A proper niche in the environment may enable a company to gain a stronger competitive advantage and avoid threats from both rivals and the environment. In order to help understand the niche of an organization, the niche concept is translated into the following specific and meaningful constructs: niche width, niche overlap and location.

The NW/O-L framework

Organizational niche width (*NW*) has been defined as the variance in resource utilization in the N-dimensional environment (Hannan and Freeman 1989).

In line with this concept, organizations pursuing strategies based on a wide range of environmental resources possess a wide niche and would be classified as generalists, whereas organizations following strategies based on a tight band of resources hold a narrow niche and are considered to be specialists. NW and its implications for organizational performance is a traditional issue in organizational studies (Boone *et al.* 2002; Dobrev *et al.* 2002; Ramirez *et al.* 2008; Sorenson *et al.* 2006). It is generally understood that a specialist will always outperform a generalist in any stable environment because the generalist must carry 'extra capacity' that sustains its ability to perform in different environments (Hannan and Freeman 1989). In contrast, in a variable environment, specialists have trouble surviving long unfavourable periods, whereas generalists do not (Baum and Shipilov 2006). Hannan and Freeman (1977) argued that the specialist maximizes its exploitation of the environment and accepts the risk of having that environment change, while the generalist accepts a lower level of exploitation in return for greater security.

Niche overlap (NO) is defined as the fraction of the focal organization's niche covered by the niche of another (Hannan *et al.* 2003). In general, organizations in two different niches have a potential for competition that is directly proportional to the extent that their organizational niche overlaps (Baum and Singh 1994). High overlap indicates that companies are substitutes or they serve the same needs and the differentiation is small, whereas low overlap indicates that different needs are served and the differentiation is great (Ramirez *et al.* 2008). NO is thus often adopted as an indicator to reflect the competition among organizations. Small values for NO imply less threat of competition among organizations and are seen as positive indicators for organizational performance. The reverse is true where NO is large.

Carroll *et al.* (2002) considered that organizational viability depends not only on NW and NO but also location (L) within an environmental space. This location is not the geographic location in new economic geography (Fujita *et al.* 2001) but the niche location in the multidimensional resource space. It assumes that resources are unevenly distributed in a multidimensional environment. The joint distribution of each dimension displays a unimodal peak representing what is called 'the market centre', where it is more bountiful or lucrative than others areas. The relatively infertile areas distributed around the market centre are designated as the peripheral area. Where L sits, relative to the market centre, is critical to an organization (Dobrev *et al.* 2002). Companies located near to the centre usually gain more market resources and opportunities. However, fierce competition in the market centre results in a high exit rate. Over time, only a few large companies reside in the centre of the market, while most companies are distributed around the peripheral area. Carroll (1985) demonstrated that generalists are more likely to be located in the centre of the market, since a position in this resource-rich area provides generalists with the potential to reap scale advantages and to grow and expand further. In contrast, specialists seem to face more competition threat and risks than their generalist competitors, as their assets may be fully exposed to the intense competition of the central location.

By integrating niche width (**NW**), niche overlap (**NO**) and location of organization (**L**), an NW/O-L analytical framework can be established. As shown in Figure 7.6, with **NW** and **L**, it is easier for contractors to understand their niche in the multidimensional international construction environment, such as their market resource utilizations and distance from the market centre. Furthermore, contractors can see whether they are in an appropriate niche. As generalists are supposed to be at the centre of the market and specialists at the periphery, Area I and Area IV seem to be more suitable niches for contractors. As a whole, this NW/O-L framework can help contractors distinguish their niche and improve their probability of sustaining it.

Similarities and differences with other theories in international construction

In a strict sense, the niche concept is not entirely new to international construction researchers. Porter's (1980) generic strategies theory, emphasizing cost leadership, differentiation and focus, is a widely adopted theory in international construction studies. In particular, differentiation is concerned with creating something that is perceived by consumers and the market as

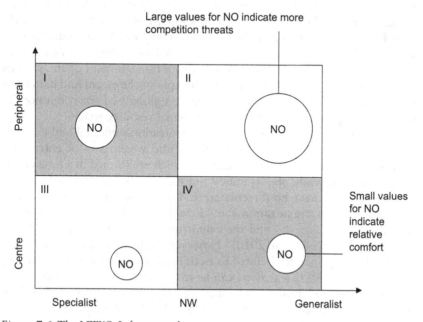

Figure 7.6 The NW/O-L framework

special, which is similar to the concept of the *NO* in niche theory. The focus implies that a company would compete in limited market segments and is related to the idea of *NW*. However, the essence of niche theory is different from Porter's theory. Niche theory is in favour of environment-driven structures for the survival of organizations. It emphasizes the natural selection process, arguing that a proper niche or environment is the primary mechanism for explaining the performance of an organization (Hannan and Freeman 1989). In contrast, Porter's generic strategies focus on strategic analysis, planning and choice and their effects on organizational performance (Korkmaz and Messner 2008). Whittington (2001) concluded that the main difference between these two theories is that Porter suggests that the performance of the companies is determined by endogenous factors such as the organizational structure, product categories, managers' decisions and the like, while niche theory is concerned more with emergent processes of natural selection, recognizing the exogenous factors (environment and the degree of fit with the environment) as the main effect on company performance.

Dunning's eclectic paradigm is one of the most important classic theories in internationalization frameworks. It can be represented by an OLI model, suggesting that the determinants of internationalization rely on the ownership (*O*), internalization (*I*) and locational (*L*) advantages that may be exploited by firms (Dunning 2000), and serves as a platform for explaining international activities, including construction. Based on this paradigm, Low and Jiang (2004) developed an OLI+S model and applied it to the international construction industry. The specialized field (*S*) that a firm is involved in and the particular country where a company is located (*L*) are closely related to the concept of *NW*. However, Dunning emphasizes comparative advantages by concentrating more on the added value that a particular field or country may offer to multinational corporations instead of the various resources utilized by them. This approach explains the extent and pattern of the foreign value-added activities of firms in a globalized sense, but not the position of organizations in a multidimensional resource space.

SWOT analysis investigates both an organization's internal and external conditions (Weihrich 1982). The strengths and weaknesses of enterprises are usually considered as internal factors, which are formed in a long development process, while opportunities and threats are external factors over which enterprises have no direct control. The philosophy behind a SWOT analysis is that an organization should establish a fit between its internal strengths and weaknesses and the opportunities and threats posed by its external environment (Lu 2010). However, with an emphasis more on resource-based principles (related to both internal and external resources), the outcome of a SWOT analysis can be complex, as it involves such things as competition and business sustainability. Furthermore, as it is mainly based on questionnaires and interviews, SWOT usually takes the form of a subjective evaluation.

Based on the four generic approaches to strategies suggested by Whittington (2001), the differences between niche theory and traditional frameworks are summarized in Figure 7.7.

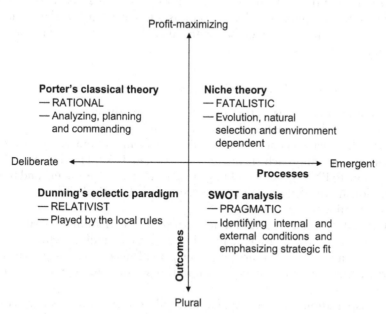

Figure 7.7 Differences between niche theory and traditional frameworks

Data

Ye *et al.* (2009) claim that it is difficult to collect data on business competition and to identify those contractors who have international operations, while Ruddock (2002) found that data on construction activities is usually poor and erratic in both domestic and international contexts. In such circumstances, ENR is a valuable resource because it provides a comprehensive historical database of international construction activities and the major actors involved (Drewer 2001). The ENR annual survey started in 1979 following the expansion of international demand for construction. It collects data from the top 225 international contractors (top 250 from the 2013 edition), including each firm's revenue and details of their submarkets, thereby offering a relatively objective and comprehensive longitudinal database for studies of international construction. Although some researchers might question the validity of the ENR data, since it is self-reported, it can be supplemented by data derived from other public sources such as company yearbooks. As most international contractors are listed companies, they are required by law to reveal data to their shareholders and maintain its integrity. Based on the ENR reports, and by comparing datasets from other sources to achieve concurrent validity, it is possible to determine the niche of international contractors and the influence of this niche on their performance. To produce a time axis, six years' data reflecting the performance of organizations from 2004 to 2009 were gathered for this purpose. This period was selected as one of relative stability given the effects of the global financial crisis did not appear in the data until 2010.

Zoning international construction markets and organizations

The international construction market represents the competitive environment in which international contractors operate. When applying niche theory, the environment should be defined from the outset by using an N-dimensional approach (Hannan and Freeman 1989; Hutchinson 1978). According to ENR, general building, industrial process/petroleum and transportation are the three most important submarkets with abundant resources for international contractors.

'General building' is a traditional market in international construction. Chiang *et al.* (2001) considered that the traditional building sector is labour intensive and does not require proprietary or advanced technology. The low entry barrier means that competition in this sector is more intense than in others.

'Transportation' markets expand fast. This is ascribed to bustling economies in developing countries and investment from both the public and private sectors. The former creates a huge demand for transportation projects and the latter finances the projects that satisfy this demand. Stimulus packages in many countries after the subprime crisis in 2008 further reinforced the transportation market.

According to ENR, the main regional markets for international construction activities are Europe, Asia and the Middle East. The African market has also witnessed a dramatic expansion since 2007.

The Middle East market fluctuates in terms of its oil production and related construction projects. As a result of a rise in oil prices and following huge expansion plans in the oil, gas and petrochemical sector, there is huge potential in this market for transportation, infrastructure, petrochemicals and water related projects.

Asia contains more than 50% of the world's population. With many developing countries and relatively high population densities, Asia has long been recognized as a market with the greatest potential for international construction activities (Raftery *et al.* 1998). The Asian market continues to play an important role in international construction, though the financial crisis in 1997 depressed this market for years. Following an international construction boom worldwide, Asia rebounded in 2003, and it is now the second-largest market in international construction.

Europe, which can be generally divided into West Europe and East Europe, is the world's biggest regional market. West Europe is a vast and stable market with modest cross-border activity, while emerging East Europe offers more opportunities. East Europe has been fuelled by building and urban infrastructure needs and foreign investment.

Africa shows huge potential construction demand, yet economic difficulties prevent this demand from being translated into projects. African international revenue began to rise in 2001, driven by North Africa (Egypt, Algeria, Nigeria and Libya). An influx of oil revenue into Africa has driven this market, making it the fastest-growing regional market in the world.

Ngowi *et al.* (2005) explained that international construction has a pattern whereby companies from advanced industrialized countries (AIC) carry out work in newly industrialized countries (NIC) or developing countries. This was supported by ENR in the 1980s, as contractors from advanced industrialized countries (i.e. those that were based in the US, Europe and Japan) dominated the international construction market. With better technology, management capacity and financial skills, these contractors were well placed to compete in the global marketplace. However, more and more developing countries, generally belonging to NIC, have joined this market. Compared with AIC contractors, the advantages of these new competitors include lower labour, material and equipment costs, advancement in certain technologies and good relationships with developing countries (Lu *et al.* 2009; Zhao and Shen 2008; Zhao *et al.* 2009), and given usually lower and aggressive bidding prices, they provide fiercer competition in the international construction market. International contractors therefore need a proper niche in order for their business to be sustainable.

Modelling the NW/O-L framework

Niche width is an important indicator that reflects organizational resource utilization. Both dimensions of product and geography are considered in the NW calculation. For the product **NW** of organization i (NW_{ip}), the definition of niche width proposed by Hannan and Freeman (1989) is adopted:

$$NW_{ip} = -\sum_{r=1}^{R} u_r \log u_r \qquad (7.1)$$

Where u_r stands for the international revenue of product r within the total international revenues of organization i. R is total number of products, including general building, manufacturing, power, water, sewer waste, industrial process/petroleum, transportation, hazardous waste and telecommunication. When the contractor is concerned with only one product, the niche width has the minimum value of zero. When the contractor's revenue is equally distributed across all the product categories, its niche width approaches the maximum value as $\log R$.

Owing to a limitation in the ENR data where revenue data based on the geography dimension is not available, equation (7.1) for NW_{ip} cannot be simply applied to geography NW of organization i (NW_{ig}). However, the NW_{ig} is identified as an important indicator to reflect the contractors' resource utilization in the geography dimension. In order to overcome the data limitation, some researchers use the span covered by the niche to reflect the resource utilization of the company. For example, Baum and Singh (1994) defined the niche of day-care centres as the span of ages that they are authorized to enrol. Dobrev *et al.* (2001) characterized the technology niche of an automobile manufacturer as the difference in sizes between the largest and smallest engines that they produce. This study defined the NW_{ig} as geographical span of organization i that they have engaged in:

$$NW_{ig} = n / N \tag{7.2}$$

where n is the number of countries in which organization i has a presence and N is the total number of countries with international construction activities.

Baum and Singh (1994) considered that NO among two organizations are, in general, asymmetric, i.e., $NO_{ij} \neq NO_{ji}$. The company with large size will exert a larger pressure on the company with small size than vice versa. Based on this hypothesis, NO_{ij} is defined as the organization i's NO with organization j, indicating the amount of competition threat that the organization i has received from organization j (Sohn 2001). It is calculated as:

$$NO_{ij} = \frac{\sum_{r=1}^{R} w_{ir} w_{jr}}{\sum_{r=1}^{R} w_{ir}^2} \tag{7.3}$$

where w_{ir} indicates the intensity of resource r used by organization i. For the product dimension, w_{ir} stands for the ratio of organization i's international revenue on product r within the total international revenue of TIC 225 for product r. Where the geography dimension is concerned, resource r denotes total project numbers[1] of TIC 225 in region r. R ($r = 1,..., 6$) that comprise North America, Europe, Latin America, Asia, the Middle East and Africa. In order to estimate the NO of an organization comprehensively, the niche overlap of organization i (NO_i) is defined as:

$$NO_i = \sum_{j=1}^{N} NO_{ij} \tag{7.4}$$

where $N = 225$, and NO_i represents the whole competitive threat that organization i has received from other companies.

Since the market environment is assumed to be unevenly distributed, location to the market centre (environment with more resources) becomes important to the company. *L* in this study is defined as the distance away from the centre of the market. The market centre must be described first. As the centre of the market is difficult to describe quantitatively, this study follows the definition by Dobrev *et al.* (2001) and assumes the largest organizations form the market centre. It thus can be defined as:

$$Centre_r = E4_r^{min} + (E4_r^{max} - E4_r^{min}) / 2 \qquad (7.5)$$

where $Centre_r$ represents centre for product/geography *r*.

For product analysis, $E4_r^{min}$ is the minimum revenue of product *r* among the top four international construction firms, while $E4_r^{max}$ is the maximum revenue of product *r* among the top four firms. For geographical analysis, $E4_r^{min}$ is the minimum project number in region *r* among the top four, and $E4_r^{max}$ is the maximum project number in region *r* among the top four. As Figure 7.8 and Figure 7.9 show, though fluctuating, these centres coincide

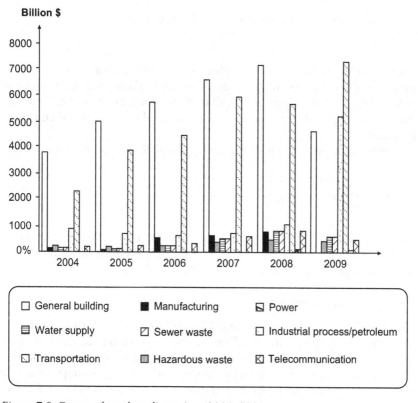

Figure 7.8 Centre of product dimension (2004–2009)

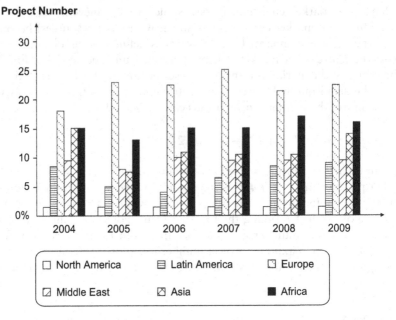

Figure 7.9 Centre of geography dimension (2004–2009)

with the main markets analysed earlier, demonstrating an asymmetric distribution of the environment resources. General building, industrial process/petroleum and transportation (product dimension), and Europe, Asia and Africa (geography dimension) are the centre of the international construction market.

L of organization i (L_i) is then calculated with Euclidean distance.

$$L_i = \sqrt{\sum_{r=1}^{R}(U_r - Centre_r)} \tag{7.6}$$

For company i's L_{ip} in the product dimension, U_r is international revenue of product r ($r = 1,\ldots, 9$), while for L_{ig} in the geographical dimension, U_r is numbers of projects in the region r ($r = 1,\ldots, 6$).

NW/O-L and performance

Good performance of an organization is usually associated with more profits, additional growth and improved market position. However, most of these indicators often lack integrity and standardization across different countries for evaluating a contractor's actual performance. Since this chapter focuses on the performance of international contractors, a project-oriented international revenue approach is chosen to measure performance. To diminish the influence of inflation and exchange rate fluctuations on the revenue of international contractors, the ranking of TIC 225 is introduced as a proxy. Though this indicator may not comprehensively reflect the performance of international contractors, it is an available and trusted indicator. Based on the ranking data of ENR, Han *et al.* (2010) investigated strategies for contractors to sustain growth in the global construction market, and Low and Jiang (2004) compared international construction performance at country level.

International contractors may choose different competitive strategies. Will contractors with different niches show different performance? In order to prove whether a contractor's niche is related to their performance, cluster analysis is used with international contractors divided into groups according to their niche.

The basic principle of cluster analysis is to classify a set of values or variables into a proper number of groups or clusters (Harrigan 1985). Since the proper number of clusters is initially unknown, a hierarchical cluster process is chosen utilizing the Ward method (Ward 1963) to estimate the numbers of clusters and their centroids for group classifications. Based on the equations (7.1) to (7.6), NW/O-L of TIC 225 in 2009 is introduced into this model. The dendrograms for both product and geography dimensions are calculated separately. With a further analysis of the two dendrograms, TIC 225 is classified into three groups based on their niche within the product dimension, while in the geography dimension there are two clusters. Tables 7.1 and 7.2 show the cluster analysis results. It can be concluded that different niche groups generally show distinguished performances (ranking), indicating that a contractor's niche is highly related to its performance in the international construction market.

In the product dimension, all 17 contractors concentrated in Cluster 1 belonged to the top 50 contractors in 2009. According to Table 7.1, these contractors controlled nearly half of the international revenue of TIC 225 in 2009, indicating their superior power and performance in the international construction market. Referring to their niche, it can be seen from Table 7.1 that they generally have a wide niche (generalists), distributed near to the market centre with a small niche overlap, and thus they mainly fall in the area IV shown in Figure 7.6. According to Table 7.3, more than 40% of their revenue comes from transportation, which was the largest market centre in 2009 (see Figure 7.7). This indicates that contractors from Cluster 1 tend to be located in the prolific resource space. Most contractors in Cluster 1 belong to AIC, such as the US, Germany, France and Spain, which supports the argument by Ngowi *et al.* (2005) that with experience in the international market, contractors from AIC are more likely to occupy market share than competitors from other countries.

Most TIC 225 companies belong to Cluster 2, contributing 53.5% of revenue to international construction in 2009. In contrast to contractors in Cluster 1, contractors in this group primarily present a narrow niche,

Table 7.1 Cluster results in the product dimension

	No. of contractors	Number Share (%)	Revenue Share (%)	Mean NWp	Mean NOp	Mean Lp	Top 1–50	Top 51–150	Top 151–225
Cluster 1	17	7.589	45.628	0.446	20.755	6712.568	17	0	0
Cluster 2	189	84.375	53.515	0.267	216.740	9757.269	33	100	56
Cluster 3	18	8.036	0.857	0.330	1142.321	10041.815	0	0	18
Total	224	100	100						

	No. of contractors	US	Japan	Korea	China	Turkey	UK	Germany	France	Italy	Spain	Others
Cluster 1	17	3	0	0	1	0	1	2	2	0	2	6
Cluster 2	189	16	13	12	47	29	2	2	2	19	8	39
Cluster 3	18	1	0	0	5	4	1	0	1	3	1	2
Total	224	20	13	12	53	33	4	4	5	22	11	47

Table 7.2 Cluster results in the geography dimension

	No. of contractors	Number Share (%)	Revenue Share (%)	Mean NWg	Mean NOg	Mean Lg	Top 1–50	Top 51–150	Top 151–225
Cluster a	78	35.455	76.318	0.197	14.744	23.330	40	33	5
Cluster b	142	64.545	23.682	0.043	55.139	31.231	9	66	67
Total	220	100	100						

	No. of contractors	US	Japan	Korea	China	Turkey	UK	Germany	France	Italy	Spain	Others
Cluster a	78	13	7	1	16	1	3	4	5	8	7	13
Cluster b	142	7	6	11	37	32	1	0	0	14	4	30
Total	220	20	13	12	53	33	4	4	5	22	11	43

are located in the peripheral area of the international construction market and show a relatively large niche overlap. It can be concluded that most contractors in Cluster 2 are consistent with the characteristics of Area I of Figure 7.6. Cluster 2 is mostly composed of contractors from China and Turkey. As new players in the international construction market, most of the contractors from NIC are still specialists operating in the peripheral area. The high entry barriers in the market centre still prevent most NIC contractors from entering.

There are 18 contractors in Cluster 3. These contractors all belong to the top 151 to 225 group and contributed less than 1% to international construction revenue in 2009, suggesting a relatively poor performance in the international construction market. Compared with the other two clusters, contractors in Cluster 3 show a mean niche width of 0.33, which is between Cluster 1 and Cluster 2. However, the average niche overlap and distance from the market centre are much larger for this group than for the other two groups, implying that they encounter greater competition threats than contractors in the other two groups. Most contractors in Cluster 3 belong to the Area II in Figure 7.6. Table 7.3 indicates that general building accounts for a significant portion of revenue for these contractors. Since the competition is more fierce in the traditional general building market (Chiang *et al.* 2001), it is understandable that the niche overlap of Cluster 3 is large.

In order to provide an intuitive understanding of international contractors' niches, the distribution of the TIC 225 for different clusters has been drawn (Figure 7.10). The horizontal axis represents the niche width of contractors, while the vertical axis shows the location in relation to the market centre. The proportion of the bubbles demonstrates the niche overlap.

As Figure 7.10 shows, contractors in Cluster 1 stay within a particular niche compared with the other two clusters. As mentioned, contractors in Cluster 1 mainly fall in Area IV of Figure 7.6. It seems that contractors within this area have relatively little competition and good performance, highlighting the proper niche for these contractors. Contractors in Cluster 2 are very different compared with those in Cluster 1. Though they do not exactly fit into Area I of Figure 7.6 as assumed previously, the general distribution of most contractors shows that they cannot enter into the

Table 7.3 Revenue share of different clusters (product dimension) (%)

	General building	Manufacturing	Power	Water supply	Sewer waste	Industrial process/ petroleum	Transportation	Hazardous waste	Telecommunication
Cluster 1	25.88	0.27	4.59	2.84	1.85	20.33	42.74	0.10	1.39
Cluster 2	20.83	1.62	13.91	3.18	1.43	38.84	19.86	0.09	0.25
Cluster 3	34.34	0.00	8.40	0.32	0.14	28.08	28.62	0.00	0.10

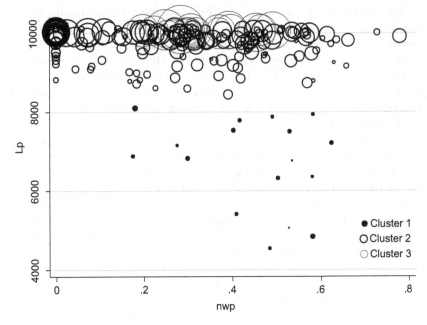

Figure 7.10 NW/O-L of TIC 225 in 2009 (product dimension)

market centre via the product dimension. Most contractors generally live in peripheral areas with a narrow niche. As most bubbles in Cluster 2 are larger than those in Cluster 1, it indicates that they encounter more competition threats than their competitors in the market centre. Contractors in Cluster 3 are mainly generalists living in peripheral areas. However, they have to bear more competition than their specialist competitors, indicating that this niche is not ideal for contractors operating in the international market.

Contractors are divided into two clusters in the geography dimension. As presented in Table 7.2, 40 out of the 78 contractors in Cluster A belonged to the top 50 contractors in 2009. Contractors from Cluster A generated nearly 80% of international construction revenue, illustrating their superior performance compared to Cluster B. Similar to contractors in Cluster 1 of the product dimension, contractors in Cluster A of the geography dimension also drop into Area IV of Figure 7.6, suggesting that this is the preferred niche for the international contractors in both the product and geography dimensions. As shown in Table 7.4, the European revenue share of contractors in Cluster A accounts for a large portion of their total revenue, and the proportion is much higher than that of their counterparts in Cluster B. As Europe was the richest market centre in 2009 (Figure 7.9), their dominant position in this market meant that they had a superior performance in

Table 7.4 Revenue share of different clusters (geography dimension) (%)

	North America	Latin America	Europe	Middle East	Asia	Africa
Cluster A	2.75	12.20	28.68	14.39	22.74	19.24
Cluster B	2.66	8.64	14.17	25.25	25.58	23.70

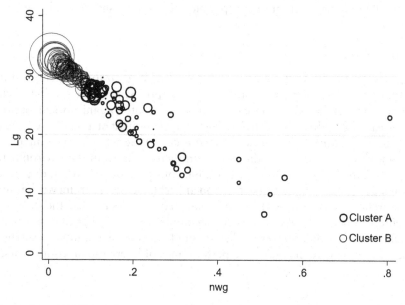

Figure 7.11 NW/O-L of TIC 225 in 2009 (geography dimension)

the international construction market. Most contractors in Cluster A come from AIC, such as the US, Japan and some European countries. The relatively similar cultural environment and geographic proximity to the European market have offered the contractors in Cluster A more advantages than their competitors, which further raises the entry barriers in this market. Meanwhile, there are 16 Chinese contractors in this group, suggesting that some contractors from NIC have benefited from the worldwide expansion process.

Contractors from Cluster B mainly have a narrow niche and a large niche overlap and are located far away from the market centre, which coincides with Area I in Figure 7.6. The majority of the members in Cluster B come from China and Turkey, implying that most contractors in NIC with a narrow niche in the geography dimension still focus on regional work rather than exploring the worldwide market. The Middle East, Asia and Africa are the main targets of Cluster B (see Figure 7.11).

It is shown that contractors in Cluster A generally show a small niche overlap, indicating their sustained ability in this competitive environment. By contrast, most contractors in Cluster B suffer from high competition in the peripheral area. Compared with the distribution of TIC 225 in the product dimension, a contractor's niche in the geography dimension is more regular. There is a significant relationship between NW_g and L_g. The contractors near the market centre are mostly generalists, while specialists are mainly scattered in the peripheral area. Most contractors choose their niche carefully and keep within the proper niche bounds as shown in Figure 7.6.

Conclusion

Cluster analysis has revealed a link between an organization's niche and its performance. However, as cluster analysis is approximate, the specific mechanism through which niche can be translated into organizational performance needs further study. Niche theory can be thought of as negative since it emphasizes natural selection without considering the activities of companies. Although a clear understanding of a contractor's status in the environment is important for the sustainable development of the organization, organizational performance is complicated, and any isolated theory framework is not comprehensive. Cheah and Wong (2004) have suggested that the different theoretical fields should be viewed as complementary rather than mutually exclusive; therefore, further studies should focus on using a more combined view to investigate the niche and performance of international contractors.

Note

1 As the exact project numbers cannot be obtained, one company in one country is assumed as one project in this study.

References and Further Reading

Baum, J. A. C. and Shipilov, A. V. (2006) Ecological approaches to organizations. In: Clegg, S., Hardy, C. and Nord, W. (eds) *Handbook of Organization Studies*, 55–110 (London: SAGE Publications).

Baum, J. A. C. and Singh, J. V. (1994) Organizational niches and the dynamics of organizational founding. *Journal of Organization Science*, 5 (4), 483–501.

Bon, R. and Crosthwaite, D. (2001) The future of international construction: some results of 1992–1999 surveys. *Building Research & Information*, 29 (3), 242–247.

Boone, C., Carroll, G. R. and Van Witteloostuijn, A. (2002) Resource distributions and market partitioning: Dutch daily newspaper, 1968 to 1994. *American Sociological Review*, 67, 408–431.

Carroll, G. R. (1985) Concentration and specialization: dynamics of niche width in populations of organizations. *American Journal of Sociology*, 90, 1262–1283.

Carroll, G. R., Dobrev, S. D. and Swaminathan, A. (2002) Organizational processes of resource partitioning. *Research in Organizational Behavior*, 24, 1–40.

Cheah, C. Y. J. and Wong, W. F. (2004) Management studies of the Chinese construction industry: which field of theories? *20th Annual Conference of the Association of Researchers in Construction Management*, Edinburgh, UK.

Chiang, Y. H., Tang, B. S. and Leung, W. Y. (2001) Market structure of the construction industry in Hong Kong. *Construction Management and Economics*, **19** (7), 675–687.

Clark, R. (1985) *Industrial Economics* (Oxford: Basil Blackwell Inc.).

Davis Langdon (2012) *World Construction 2012* (London: Davis Langdon).

Dimmick, J., Chen, Y., and Li, Z. (2004) Competition between the Internet and traditional news media: the gratification-opportunities niche dimension. *Journal of Media Economics*, **17** (1), 19–33.

Dobrev, S. D., Kim, T. Y. and Carroll, G. R. (2002) The evolution of organizational niches: U.S. automobile manufacturers, 1885–1981. *Administrative Science Quarterly*, **47** (2), 233–264.

Dobrev, S. D., Kim, T. Y. and Hannan, M. T. (2001) Dynamics of niche width and resource partitioning. *American Journal of Sociology*, **10** (5), 1299–1337.

Drewer, S. (2001) A perspective of the international construction system. *Habitat International*, **25** (1), 69–79.

Dunning, J. H. (2000) The eclectic paradigm as an envelope for economic and business theories of MNE activity. *International Business Review*, **9** (2), 163–190.

ENR (2012) The top 225 international contractors. *Engineering News Record*, **267** (8), 1–21 (compiled by Reina, P. and Tulacz, G. J.).

Fujita, M., Krugman, P. R. and Venables, A. J. (2001) *The Spatial Economy: Cities, Regions, and International Trade* (Boston: Massachusetts Institute of Technology).

Han, S. H., Kim, D. Y., Jang, H. S. and Choi, S. (2010) Strategies for contractors to sustain growth in the global construction market. *Habitat International*, **34** (1), 1–10.

Hannan, M. T., Carroll, G. R. and Polos, L. (2003) The organizational niche. *Sociological Theory*, **21** (4), 309–340.

Hannan, M. T. and Freeman, J. (1977) The population ecology of organizations. *American Journal of Sociology*, (82), 924–964.

Hannan, M. T. and Freeman, J. (1989) *Organizational Ecology* (London: Harvard University Press).

Harrigan, K. R. (1985) An application of clustering for strategic group analysis. *Strategic Management Journal*, **6** (1), 55–73.

Hutchinson, G. E. (1978) *An Introduction to Population Ecology* (New Haven: Yale University Press).

Kale, S. and Arditi, D. (2002) Competitive positioning in Untied States construction industry. *Journal of Construction Engineering and Management*, **128** (3), 238–247.

Korkmaz, S. and Messner, J. I. (2008) Competitive positioning and continuity of construction firms in international markets. *Journal of Management in Engineering*, **24** (4), 207–216.

Low, S. P. and Jiang, H. B. (2004) Estimation of international construction performance: analysis at the country level. *Construction Management and Economics*, **22** (3), 277–289.

Low, S. P., Jiang, H. B. and Leong, C. H. Y. (2004) A comparative study of top British and Chinese international contractors in the global market. *Construction Management and Economics*, **22** (7), 717–731.

Lu, W.S. (2010) Improved SWOT approach for conducting strategic planning in the construction industry. *Journal of Construction Engineering and Management*, **136** (12), 1317–1328.

Lu, W.S., Li, H., Shen, L.Y. and Huang, T. (2009) Strengths, weaknesses, opportunities, and threats analysis of Chinese construction companies in the global market. *Journal of Management in Engineering*, **25** (4), 166–176.

Moore, J.F. (1996) *The Death of Competition: Leadership and Strategy in the Age of Business Ecosystems* (New York: Harper Collins).

Ngowi, A., Pienaar, E., Talukhaba, A. and Mbachu, J. (2005) The globalization of the construction industry – a review. *Building and Environment*, **40** (1), 135–141.

Ofori, G. (2003) Frameworks for analysing international construction. *Construction Management and Economics*, **21** (4), 379–391.

Öz, Ö. (2001) Sources of competitive advantage of Turkish construction companies in international markets. *Construction Management and Economics*, **19** (2), 135–144.

Porter, M.E. (1980) *Competitive Strategy: Techniques for Analyzing Industries and Competitors* (New York: Free Press).

Raftery, J., Pasadilla, B., Chiang, Y., Hui, E. and Tang, B. (1998) Globalization and construction industry development: implications of recent development in the construction sector in Asia. *Construction Management and Economics*, **16** (6), 729–737.

Ramirez, A., Dimmick, J., Feaster, J. and Lin, S.F. (2008) Revisiting interpersonal media competition – the gratification niches of instant messaging, e-mail and the telephone. *Communication Research*, **35** (4), 529–547.

Ruddock, L. (2002) Measuring the global construction industry: improving the quality of data. *Construction Management and Economics*, **20** (7), 553–556.

Sohn, M.W. (2001) Distance and cosine measures of niche overlap. *Social Networks*, **23** (2), 141–165.

Sorenson, O., McEvily, S., Ren, C.R. and Roy, R. (2006) Niche width revisited: organizational scope, behavior and performance. *Strategic Management Journal*, (27), 915–936.

Tisdell, C. and Seidl, I. (2004) Niches and economic competition: implications for economic efficiency, growth and diversity. *Structural Change and Economic Dynamics*, **15**, 119–135.

Ward, J.H. (1963) Hierarchical grouping to optimize an objective function. *Journal of the American Statistical Association*, **58** (301), 236–244.

Weihrich, H. (1982) The TOWS matrix: A tool for situational analysis. *Long Range Planning*, **15** (2), 54–66.

Whittington, R. (2001) *What Is Strategy – and Does It Matter?* (London: Thomson Learning).

Ye, K.H., Lu, W.S. and Jiang, W.Y. (2009) Concentration in the international construction market. *Construction Management and Economics*, **27** (12), 1197–1207.

Zhao, Z.Y. and Shen, L.Y. (2008) Are Chinese contractors competitive in international markets? *Construction Management and Economics*, **26** (3), 225–236.

Zhao, Z.Y., Shen, L.Y. and Zuo, J. (2009) Performance and strategy of Chinese contractors in the international market. *Journal of Construction Engineering and Management*, **135** (2), 108–118.

Editorial comment

In 1974, the US–based Xerox Company had around 85% of the photo-copier market, but that figure had dropped to less than 20% by 1984. The dramatic drop was largely due to the rise of competitors such as Ricoh and Canon, which were producing similar machines for about half the cost. In an effort to recover lost market share, Xerox instituted a comprehensive system of *benchmarking*. This involved a careful and detailed analysis of how the company operated and in particular of the processes associated with the manufacture of their products.

Benchmarking in various forms is now a well-established and widely used technique for monitoring and improving business performance. Basically it involves identifying best practice in an industry and then seeking ways to improve performance in order to match or exceed best practice.

Measuring performance is fundamental to the benchmarking process, but in construction there is little agreement on how performance should be measured or, indeed, on precisely what should be measured. Some studies look at *productivity*, others at *efficiency* or *effectiveness*. Some combination of these may be described as *performance*, but there is no clear definition for that term that applies perfectly in the context of construction. The problem is exacerbated by the difficulties associated with measuring construction quality.

Construction industry productivity may be measured at several levels, from that of site activities such as bricklaying to macroeconomic measures of national industries based on hours worked and value added. In the following chapter, a detailed method for measuring comparative performance at an intermediate level of individual projects is described. It utilizes basic project data that is often readily available from Internet sources. The reduction in dependence on data that is typically very difficult to collect from industry is one of the strengths of the method. Another advantage is that it is based on data for real completed projects rather than estimates for hypothetical projects. In its original form, it has been applied in several studies of project performance over a number of years, and here the method is refined and expanded to include assessment of quality and cost efficiency.

The method provides a framework for comparisons and thus benchmarking at a number of levels, which range from a comparison of as few as two projects to comparisons of national construction industries based on data from large numbers of completed projects. Relative performance can be assessed across a number of firms and/or a number of locations such as cities or regions. The method is explained here in terms of the underpinning theory and illustrated by means of case studies based on extensive datasets from the US and Australia.

8 Performance measures for construction

Craig Langston

Introduction

The global construction industry is highly competitive, fragmented and cyclical and frequently operates on low margins (Loosemore 2003). Yet construction accounts for a significant portion of economic activity and is a catalyst for many other sectors. The industry is also labour intensive and project specific and involves team relationships that form and disband on a regular basis. It is not surprising, therefore, that construction performance and reform have dominated research within the industry for more than 50 years.

Yang *et al.* (2010) undertook a critical literature review of performance measurement in construction. Their work provided an excellent platform from which to propose a fresh approach to the problem. They classified performance measurement studies into three categories: project, organizational and stakeholder. The major frameworks were shown to be the European Foundation for Quality Management excellence model, key performance indicators (KPIs) and balanced scorecards. Gap analysis (e.g. trend analysis), integrated performance index (e.g. multiple criteria analysis), statistical methods (e.g. regression analysis) and linear programming (e.g. data envelopment analysis) were shown to be the most frequently applied research methods for performance measurement.

Performance measures are approaches to determine if a process has obtained the desired result. However, the diversity of the construction process makes it difficult to apply just a simple definition. In reality, performance is relative and assessed via comparison to observed best practice. This requires appropriate and current data in an objective (i.e. numeric) format across a wide range of building types, locations, times and regulatory environments that makes the task difficult if not impossible to complete.

The construction industry has long been criticized for apparent underperformance (e.g. Pieper 1989). Reports such as those from Latham (1994) and Egan (1998) have called for a rethink of the traditional construction process, but more than a decade on, many might conclude that little has changed. So the debate continues, and the search for appropriate measures lies on the leading edge of research into the performance of contractors, projects and industries and probably will do so well into the future.

Yang *et al.* (2010:281) concluded that no single framework or approach fits all situations – all have their advantages and disadvantages – and therefore "*it is an important task to develop a more comprehensive performance measurement framework in construction in the future*". The aim here is to propose a new model for performance measurement and to test it using what is understood to be one of the largest samples of construction project data ever assembled across two sample countries: Australia and the United States. The analysis of this data not only demonstrates the practical application of the model but also provides new insight into the efficiency of the construction industry in these two countries over the last decade.

Performance and productivity

'Performance' and 'productivity' are often used interchangeably in the literature. To some extent, the difference lies in the type of study being undertaken and its scope. For example, studies into the merits of a single project or contractor may be best described as performance measures, even though the productivity of employed labour may be incorporated in the analysis. On the other hand, studies into the efficiency of multiple projects or contractors may help to understand industry performance, and these types of studies tend to focus on comparative productivity. In both cases, ratios of output over input are typically involved.

Benchmarking is also a common strategy. Liao *et al.* (2012) stated that the benchmarking of engineering productivity can assist in the identification of inefficiencies and thus can be critical to cost control. Their study developed a standardized approach using 'z-scores' to aggregate engineering productivity measurement from data collected from 112 actual projects and resulted in a metric incorporating a project-level view of engineering productivity. The metric enables benchmarking of heavy industrial project productivity as a basis for comparison of individual project performance. Mohamed (1996) earlier urged organizations to be actively involved in project benchmarking to assess their performance, measure their productivity rates and validate their cost-estimation databases.

Motwani *et al.* (1995) discussed the importance of measuring productivity over time. Changes in productivity rather than absolute values were seen as critical if building contractors were to be competitive and successful in the increasingly global construction market. Furthermore, Yates and Guhathakurta (1993) looked at international labour productivity differences and concluded that labour quality, motivation and management were the main issues. Labour quality, for example, may include union agreements, restrictive work practices, absenteeism, turnover, delays, availability, level of skilled artisans, use of equipment and weather. Mohamed and Srinavin (2002) found that productivity falls when thermal comfort moves away from the optimum range. Disruption was also shown to be correlated with poor management and led to low productivity (Enshassi *et al.* 2007).

Various studies have attempted to measure labour productivity at the project or task level through cost and quality management maturity (e.g. Willis and Rankin 2012), concurrent engineering (e.g. Shouke *et al.* 2010), organizational analysis (e.g. Sahay 2005), process improvement (e.g. Stewart and Spencer 2006) and human resource management (e.g. Hewage and Ruwanpura 2006).

Key performance indicators

Performance is not just about efficiency but about achieving desired results. To help in this endeavour, a wide variety of KPIs have been identified and used to measure the success of construction projects. These include indicators of client satisfaction, stakeholder engagement, service delivery, investment return, urban renewal, defect minimization, trust, dispute avoidance, innovation, safety and standard. Three of the most commonly cited KPIs are on-time completion (time), within agreed budget (cost) and nondefective workmanship as specified (quality).

Time, cost and quality necessarily interact. It is well understood in the industry and in the literature that trade-offs occur between optimizing performance for any of these KPIs. For example, accelerating completion of a project will usually involve extra cost, reducing cost will tend to lower quality and increasing quality standards will take more time to deliver.

Meng (2012:188) found that construction projects often suffer from poor performance in terms of time delays, cost overruns and quality defects and, from an analysis of previous research findings, concluded that *"time, cost and quality are the three most important indicators to measure construction project performance"*. From a survey of 400 construction practitioners in the UK, with a response rate of 30%, his research found that 35.6% of projects studied were delayed, 25.2% were overspent and 17.7% had significant defects. These problems were more prevalent in traditional procurement relationships compared to partnering (or relationship management) arrangements, and the deterioration of supply chains was a major reason for the occurrence of poor performance.

Rankin *et al.* (2008) identified a number of performance metrics suitable for KPI–style evaluation. These were divided into time, cost, quality, scope, safety and sustainability. Brown and Adams (2000) undertook 15 case studies derived from UK data and found that project management as implemented in the UK failed to perform as expected in relation to the three predominant performance evaluation criteria of time, cost and quality. In fact, they showed that project management had little effect on time performance, no effect on cost performance and a strong yet negative effect on quality performance. Other factors were assumed to be at play.

Time performance

Time performance usually means the project is completed on or before the agreed handover date. Sometimes contractual documents refer to time being the 'essence of the contract', which exemplifies the criticality of timely completion due to subsequent plans that cannot be delayed.

Time on construction projects can be measured in days, weeks or months. Obviously large projects take more time to construct than small projects, so a reasonable KPI might be square metres of gross floor area completed per month (m^2/month). This is an output measure describing production. A high value for this KPI would mean that the construction process was fast and vice versa.

Walker (1995) found four factors that significantly affect construction time performance. These comprised and can be summarized as:

1. construction management effectiveness (i.e. competence);
2. the sophistication of the client and the client's representative in terms of creating and maintaining positive project team relationships with the construction management and design team (i.e. teamwork);
3. design team effectiveness in communicating with construction management and client's representative teams (i.e. communication) and
4. a small number of factors describing project scope and complexity (i.e. work definition).

Prediction of construction time has been studied at length. One of the earlier studies was Bromilow (1969), in which a predictive model was developed using the relationship between cost and duration. It was found that the time taken to construct a project is highly correlated only with the project's size, as measured by its final cost. Love *et al.* (2005) proposed an alternative model to Bromilow and concluded that gross floor area and the number of storeys were superior determinants of time performance in forecasting construction project duration. Lin *et al.* (2011) reviewed a number of attempts at duration prediction in various countries based on the relationship between variables. Likewise, in their own research concerning steel-reinforced concrete buildings in Taiwan, regression was the chosen methodology, and cost, floor area and number of storeys were the variables showing the strongest correlations, with no significant multicollinearity detected. Change orders and rainy days were added and slightly improved the predictive reliability of their model.

Cost performance

Cost performance is normally judged relative to an agreed budget. Sometimes projects may have no budget, which means that cost is not a consideration, but this is rare. Completion close to budget is usually preferable; in some cases being well below budget is seen as an advantage, although often not. Clients tend not to like surprises, so the final project cost should be the result of prudent cost management processes and therefore, by definition, deliver an end result close to the agreed budget.

Construction cost is measured in pure financial terms, usually in local currency, and should focus on the building rather than the land (i.e. should exclude site purchase costs). Since construction often spans many years, it is necessary to bring costs to a common date. The conversion to a common date is undertaken using building price indices that reflect inflationary change appropriate to the current level of construction intensity.

Cost conversion is also required to take account of geographic location. This applies to cities or centres in a particular country or in other countries. The latter will involve different currencies and the problem of exchange rates. One solution is to establish a 'locality index' for major cities that uses the principle of purchasing power parity (PPP). For example, by pricing a representative basket of construction-related items covering labour, material and plant, a standard basket price in each city (in local currency terms) can be computed and act as a locality index. Thereafter, the cost of a project can be divided by the cost of the representative basket to obtain the equivalent number of baskets required to pay for the construction. Although the unit of measure is 'baskets', not currency, the answer is an indicator of cost performance that has no locational boundaries. For example, if Project A in Hong Kong was 5 baskets/m² and Project B in New Delhi was 4 baskets/m², then the construction cost in Hong Kong would be 25% more than that in New Delhi.

In this research, the representative basket for a city is called a *citiBLOC* (BLOC = basket of locally obtained commodities). The construction cost of a project, therefore, can be measured in citiBLOCs that will take account of location and are converted for time. The unit of cost performance employed in this study is baskets per square metre (citiBLOCs/m²). This is an input measure describing resources. Costs should ideally exclude site works since they are not proportional to building area. A high value for this KPI would mean that the construction process was expensive (i.e. either high quality, complex or inefficient) and vice versa.

Chau (1993) demonstrated that construction productivity could be measured from an analysis of only cost and price data and the relative value shares of inputs. This data is in general more readily available than the detailed information required by other methods, and therefore his proposed approach was less restrictive. A ratio of output to multiple inputs was used in his model, and all were expressed in monetary terms.

Quality performance

Quality performance is referenced to the standard of the delivered project and that specified in the contract documents. The expectation is to receive what is specified, no more and no less, and often this is judged in the detail of the finishes and the workmanship applied. There is no convenient unit of measurement for quality, and it therefore involves a collection of issues, some of which are objective (e.g. number of identified defects) and others that are subjective (e.g. craftsmanship).

Quality is influenced by a number of related factors, all of which would normally add cost and time to some extent as the level of quality increases. These include buildability, innovation, building height, extent of fit-out, environmental performance, compliance, standard of finish, supervision levels and efficiency.

While quality defies objective measurement, relative comparison is possible. Hotels, for example, are classified according to quality and assigned a star rating, so what to expect from a five-star hotel is well understood. Relative quality performance involves comparing like with like. Standards and expectations differ among residential, commercial and industrial applications, between urban and rural settings, among different countries and cultures, and among project stakeholders.

Hsieh and Forster (2006) found that the structural quality of residential construction in Taiwan fell, and fell measurably, as production reached higher levels and skilled labour shortages arose. Also in Taiwan, Yang (2009) found that the quality of project deliverables was significantly associated with automation technology usage in the front-end, design, procurement and construction phases. Furthermore, using a case study methodology, Tchidi *et al.* (2012) found that prefabrication improved project quality and reduced construction waste.

Quality management has grown in prominence over the years. Munro-Faure and Malcolm (1992) believed that quality was probably the best strategy to ensure customer loyalty, defend against foreign competition and secure continuous growth and profits in difficult market conditions. ISO-9000 (quality management) is the current international standard for assuring quality, and ISO certification is a demonstrable way that construction contractors can communicate to customers that they have systems in place to deliver quality outcomes and grow in abilities through continuous process improvement (Ali and Rahmat 2010).

Rosenfeld (2009) investigated the cost of quality via case studies and concluded that the optimal range for investment in quality is between 2% and 4% of a construction company's revenue per annum. Investing less than 2% in prevention and appraisal will definitely entail higher failure costs, whereas an investment of more than 4% most probably will not pay itself back. He also found that quality failures bear substantial hidden costs that cannot be readily measured. Quality failure is the result of not doing things right the first time (Abdelsalam and Gad 2009).

An integrated time-cost-quality performance index

Langston and Best (2001) developed a performance index (PI) based on a ratio of output (production capacity) to input (resource consumption). Production capacity was measured as constructed floor area completed per month, computed as the gross floor area (m²) divided by the time between commencement and handover (months). Resource consumption was measured as construction cost per square metre, computed as the number of representative baskets (e.g. citiBLOCs converted to a base year) divided by the gross floor area (m²). They suggested that for projects of similar quality, the resultant index produced an indicator of construction efficiency, where the higher the index the more effective was the process of the project's construction. Equation 8.1 describes the PI:

$$\text{Performance index } (PI) = \frac{production\ capacity}{resource\ consumption}$$
$$= \frac{m^2\ /\ month}{\cos t\ /\ m^2} \tag{8.1}$$
$$= \frac{a^2}{ct}$$

where: *a = gross floor area in square metres*
c = completed project cost (e.g. number of citiBLOC baskets)
t = time for completion in months

While high PI scores identify projects with strong production capacity per unit of input (i.e. construction efficiency), low PI scores conversely identify projects with strong resource consumption per unit of output (i.e. construction quality). Both can be considered advantageous. The best projects are arguably the ones that display efficiency and quality and hence are more likely to have scores around the mean.

The PI is a rare attempt to integrate both time and cost into a single performance indicator. But while it can separate projects according to their strengths, it does not clearly identify best practice. Adjustment of data to 'normalize' for either efficiency or quality/complexity is problematic. A different approach is needed.

Multiplying production capacity and resource consumption together computes a weighted measure of performance that does not disadvantage projects that are expensive on the basis of their quality and/or complexity. This alternative approach is shown by Equation 8.2:

$$Construction\ efficiency\ (CE) = production\ capacity$$
$$\times resource\ consumption$$
$$= m^2\ /\ month \times cost\ /\ m^2 \qquad (8.2)$$
$$= \frac{c}{t}$$

where: c = completed project cost (e.g. number of citiBLOC baskets)
t = time for completion in months

Given that quality and/or complexity would be expected to affect time and cost in the same direction (i.e. higher levels of difficulty take longer to build and cost more), a ratio of cost over time would see quality/complexity (and project scale for that matter) cancelled out between the numerator and the denominator. Project cost is treated, in this case, as an output measure despite being determined from the amount of money spent to construct the building (an input measure for PI). The time to construct is now the input measure in delivering the project. This approach is similar to that used by Chau (1993).

Construction efficiency (CE) is a valuable indicator for judging the efficiency of a contractor, the efficiency of construction in a particular location or the efficiency of the construction industry overall. As costs are expressed in citiBLOC terms, this analysis can be performed among projects constructed in any location, nationally or internationally. However, CE is affected by working hours/month and delays due to inclement weather, so comparing it to the mean efficiency of other projects of similar context and expressing the outcome as a ratio would be necessary to enable proper interpretation.

Chiang *et al.* (2012) investigated construction efficiency (they used the term 'productive efficiency') for contractors based in mainland China and Hong Kong. Their study reviewed single-factor productivity, defined as relating to individual contractors and focused on average labour productivity (i.e. output per hour worked), and total factor productivity, defined as relating to the entire industry but hampered by the complexity of measurement methods and the lack of available data. They found that generally both mainland China and Hong Kong contractors improved their efficiency over the period 2004 to 2010, with Hong Kong firms doing better due to their managerial rather than technical competence.

CE can be separated from overall performance to determine construction complexity (CC), defined as a mix of a project's quality standard and its buildability. Equation 8.3 shows CC is related closely to project cost per square metre (cost/m^2), and therefore the amount of resources consumed in the construction process underpins the calculation:

$$\text{Construction complexity } (CC) = \frac{CE}{PI}$$

$$= \frac{c}{t} \cdot \frac{ct}{a^2} \qquad\qquad (8.3)$$

$$= \frac{c^2}{a^2}$$

where: c = completed project cost (e.g. number of citiBLOC baskets)
a = gross floor area in square metres

Understanding project performance and determining best practice, from a stakeholder's perspective, can then be determined by a study of the factors affecting CC. These were stated earlier as including but not limited to buildability, innovation, building height, extent of fit-out, environmental performance, compliance, standard of finish, supervision levels – but not efficiency, as this is now assessed independently. As with CE, expressing the outcome as a ratio relative to the mean of a number of projects would aid interpretation.

Low (2001) compiled data from various sources to measure the relationship among buildability, structural quality and productivity in the Singaporean construction industry. Buildability was measured using the Building Design Appraisal System (BDAS) produced by Singapore's Building and Construction Authority (BCA), quality was measured by the Construction Quality Assessment System (CONQUAS) also produced by the BCA and productivity was measured as floor area completed per man-day for a range of case studies reported in the literature. The latter, however, did not seem to take account of different skill levels and would have been difficult to collect without detailed records of working patterns and rosters. Nevertheless, he found a weak correlation between buildability and quality but a stronger correlation between buildability and productivity. This sounds intuitively correct. Yet the robustness of data particularly related to average labour productivity remains an area of concern (Chang 1991; Chan and Kaka 2007; Doloi 2007; Allan *et al.* 2010).

Best practice may lie where projects have balanced scores for both CE and CC. Multiplying CE and CC scores together (i.e. c^3/a^2t) can be useful to highlight such projects. However, high cost/m^2 can be a sign of high standard or poor execution. The technique of data envelopment analysis may be a more appropriate method to assess the impact of multiple performance measures and to determine the best practice 'frontier' (e.g. Chiang *et al.* 2012; Horta *et al.* 2012).

Method

Crawford and Vogl (2006) called for further research into construction performance to focus on creating new or improving existing datasets. Advancements in this area are hampered by data quality and availability. Information is often commercial-in-confidence and powerful to those who have access, so data sharing is limited. There is no existing database in which all the relevant information can be found to undertake a robust analysis of construction performance. Nevertheless, the information exists in fragmented forms and requires considerable time to collect into a single place and fill the numerous gaps. As stated by Wegelius-Lehtonen (2001:115) and undoubtedly many before him, *"if you want to improve something – measure it"*.

A case study method is employed here to demonstrate the application of performance measurement to construction and to make international comparisons. Buildings completed in the last 10 years (2003–2012) of 20 storeys or more in height are selected as the population for the study. These projects are sourced from the Skyscraper website (Skyscraper 2012). All such projects in the five largest cities in both Australia and the United States are identified and assembled into a database. Information about these projects can be found in the public domain via the Internet, since tall buildings get publicity, but where key information is not discoverable, it can be followed up through contact with the project architect or building contractor where possible. The database fields are shown in Table 8.1.

Table 8.1 Performance index database fields

Field	Comments
Building ID	Skyscraper database identifier
Project Name	
Detailed Address	
City	
Country	
Building Type	commercial, residential, hospital, hotel, mixed, other
Date Completed	year (2003–2012)
Storeys Above Ground	number of upper floors (minimum 20)
Storeys Below Ground	number of basement floors
Building Height	roof height above ground (metres)
Area	gross floor area (square metres)
Construction Cost	A$ or US$ (excluding design, land and site costs*)
Fitout	fully furnished or shell
Construction Time	commencement to handover (months)
Architect Name	
Contractor Name	
Special Factors	unusual features noted

*External works is not very significant for high-rise buildings and is not deducted from cost in this study

A similar approach was undertaken by Kaming *et al.* (1998) using 31 high-rise projects of various types in Indonesia and by Langston and Best (2001) using 76 high-rise projects across 12 countries. Data was based on case studies with participant interviews or questionnaires. Area, construction duration and cost were collected in both cases, although the Indonesian study did not involve the complexity of multiple currencies.

The cost of the representative citiBLOC basket is priced locally for each city (mid-year 2012). The citiBLOC basket contains 10 common construction items plus an indicator of market conditions, as listed in Table 8.2. Items are of about equal value and divided into material (50%), labour (40%) and plant (10%), reflecting typical composition. A citiBLOC can act as a building price index to convert historical costs to present-day terms (actual price deflators and market conditions are applied in this study, as historical citiBLOCs are not available – see RBA 2012 and US Inflation Calculator 2012). Online data from Cordell (Australia) and RS Means (United States) are used to price the citiBLOC items consistently.

Table 8.2 Representative construction items for citiBLOC (Sydney)

Item	Standard Description	Unit	Quantity (weighting)	Local Currency (ex-tax)
Material *(supply only including CBD delivery)*				
A	32 MPa ready-mixed concrete (1 m^3 = 35.31 cu. feet)	m^3	45	11,144
B	Steel in 250 × 25.7kg/m 'I' beam (17.3 lb/foot)	t	6.8	9,350
C	10mm clear tempered glass (1 m^2 = 10.76 sq. feet)	m^2	44	10,472
D	13mm thick gypsum plasterboard (½" thick)	m^2	1,300	10,140
E	100 × 50mm sawn softwood stud (1 m = 3.28 feet)	m	2,750	9,873
Labour *(charge-out rate including on-costs)*				
F	Electrician	hr	150	9,900
G	Carpenter	hr	185	10,915
H	Painter	hr	200	10,400
I	Unskilled labour	hr	275	10,863
Plant *(third-party hire rate including operator and fuel)*				
J	50 t mobile crane	day	5	10,200
	average price per item (i.e. 1 citiBLOC):			10,326

SYDNEY, AUSTRALIA (2012)
Current Market Conditions: ☑ very competitive (low profit) ☐ normal ☐ overheated (high profit)

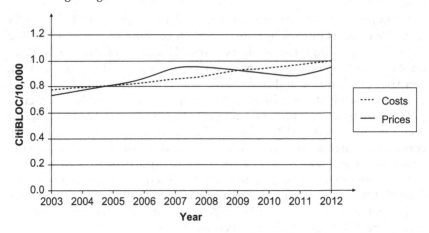

Figure 8.1 Variation in mean citiBLOC costs and prices for Australia (2003–2012)

Figure 8.1 shows an example of the citiBLOC index (avoiding and embracing current market conditions) over the study period. The former is a cost index that essentially follows the rate of inflation or deflation, while the latter is a price index that additionally considers the competitiveness of the construction marketplace and therefore takes account of wider economic factors impacting contractor profit margins (as evidenced in bids to win work). The citiBLOC cost index therefore requires some adjustment (maybe ± 10%) to reflect market prices. It is incorrect to compare citiBLOCs across countries or regions, as they are expressed in different currencies.

Case study: Australia

A total of 150 projects were identified in the five largest cities in Australia, comprising 28 in Sydney, 40 in Melbourne, 63 in Brisbane including the nearby Gold Coast (GC), 18 in Perth and 1 in Adelaide. Complete information was discovered for 86 projects, representing 57% of the known population. To achieve this conversion rate, considerable effort was made to contact project stakeholders and obtain missing data. In this process, 29% of the dataset was confirmed and validated, including the single project in Adelaide.

Table 8.3 summarizes mean performance measures for selected Australian projects. Cities are listed in descending order of size (based on population). Each measure is expressed in relation to the national mean for better comprehension, highlighting under- or overachievement, and provides a contextual frame to assess impacts such as the relative advantage of different contractors operating in various markets, local custom and climatic conditions. Note that real construction cost per square metre (expressed in

Table 8.3 Mean performance measures (selected Australian projects)

City	Projects	citiBLOC		Mean					Efficiency
		Index	m²/month	cost/m²	PI	CE	CC	Percent
Sydney	15	10,326	1,541	0.46	4,403	694	0.27	+4.11%
Melbourne	26	9,754	2,034	0.30	7,926	587	0.10	−11.94%
Brisbane/GC	38	10,294	1,587	0.47	4,879	699	0.28	+4.86%
Perth	6	9,837	1,752	0.51	3,993	860	0.28	+29.01%
Adelaide	1	9,788	1,240	0.40	3,116	493	0.16	−26.04%
Australia	Mean	10,000	1,631	0.43	4,863	667	0.22	0.00%
	CoV	2.85%	17.88%	19.08%	37.64%	20.63%	38.16%	

2012 prices) can be determined by multiplying the cost/m² column by the citiBLOC index column.

The coefficient of variation (CoV) for the citiBLOC index is extremely low, indicating that prices around Australia are quite consistent. Low CoVs (i.e. less than 30%) are also evident for most performance measures. However, a higher dispersion around the mean occurs for PI and CC. In the case of PI and CC, this may be explained by the broad range of factors that are included in these ratios and which are not necessarily correlated with each other.

Perth displays the highest construction efficiency (29.01% above the national average). Sydney has the highest cost base, while all cities other than Melbourne and Adelaide have robust construction complexity scores. Melbourne has the fastest production capacity and the highest performance index and appears to be the cheapest location to build in, but is 11.94% below the national average in terms of construction efficiency and has the lowest construction complexity score. It may be appropriate to discount the outcomes for Perth and particularly Adelaide on the basis of the small number of projects included in their calculation. When focusing on construction efficiency solely along the eastern seaboard of Australia, therefore, it is evident that Sydney and Brisbane/GC are comfortably outperforming Melbourne. Nevertheless, Melbourne is clearly cost competitive.

Figure 8.2 shows the distribution of project types, with approximately half designated as residential use. Most projects have fit-out included in their construction cost. Mean citiBLOC cost/m² is shown in brackets.

Figure 8.3 displays values for construction efficiency compared with cost/m² over time, based on all 86 Australian projects according to their year of completion. A rise in cost/m² is evident, but a greater rise in efficiency means that 'real' construction efficiency has relatively grown between 2003 and 2012. The cost/m² reflects increase in building quality and/or building complexity, which is logical given rising living standards, although an alternative explanation perhaps is an inappropriate cost-adjustment algorithm.

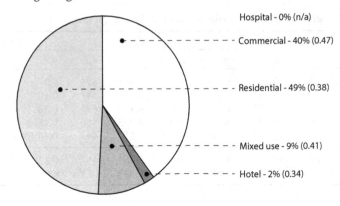

Figure 8.2 Project types (selected Australian projects)

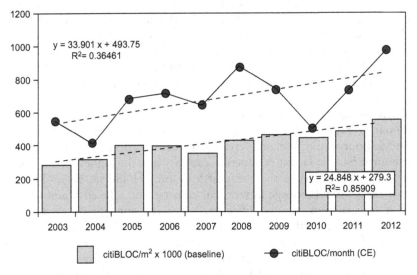

Figure 8.3 Construction efficiency over time (selected Australian projects)

Chiang *et al.* (2012) found that construction efficiency increased in both mainland China and Hong Kong over the period 2004 to 2010, despite the backdrop of a general lack of consensus that the productivity of the construction industry (globally) was improving. The results of this research support the findings of Chiang *et al.* (2012) by providing some evidence that construction efficiency has improved in Australia over the same period. Using linear regression to determine the trend, the growth in CE over the past decade is +6.87% per annum (i.e. 33.901 / 493.75 × 100). Allowing for the knowledge that base costs have grown in real terms by +5.03% per annum (i.e. 24.848 / 493.75 × 100), real construction efficiency is computed at +1.84% per annum.

Case study: United States

A total of 354 projects were identified in the five largest cities in the United States, comprising 194 in New York, 11 in Los Angeles, 113 in Chicago, 25 in Houston and 11 in Philadelphia. Complete information was discovered for 251 projects, representing 71% of the known population. None of the projects have been independently validated. Table 8.4 summarizes mean performance measures for selected American projects. Cities are again listed in descending order of size.

The first point to note is that the citiBLOC basket contains a good deal of variability (i.e. the price of the standard construction basket in New York is more than 50% more than in Houston), underlining the importance of a city-based locality index rather than a national average. Each column has a low CoV and hence low dispersion around the mean, except for PI and CC. The large number of projects located in New York and Chicago adds confidence to their results, since any individual project can exert little influence on the overall mean. It is interesting, therefore, that Chicago has a performance index more than three times that of New York despite displaying equivalent levels of efficiency.

Houston demonstrates that its projects are more efficient (i.e. 33.21% above the national average). Houston has both a low cost base and a high output capacity that leads to the second-highest performance index of the group. The strength of its performance appears more a function of construction efficiency than construction complexity, although the latter is respectable.

New York attracts a premium on cost and has the slowest output rate in the study, handing it the lowest performance index by a considerable margin. Overall, it appears that complexity is high while efficiency is relatively low (9.60% below the national average). Chicago, on the other hand, has low-cost projects despite not being a cheap place to build, and both

Table 8.4 Mean performance measures (selected US projects)

City	Projects	citiBLOC Index	m²/month	cost/m²	PI	CE	CC	Efficiency Percent
					Mean			
New York	140	10,693	1,287	0.57	2963	680	0.40	−9.60%
Los Angeles	8	8,559	2,090	0.44	6,229	826	0.24	+9.81%
Chicago	71	9,026	2,300	0.30	9,275	672	0.11	−10.66%
Houston	21	7,005	2,549	0.41	8,137	1,002	0.20	+33.21%
Philadelphia	11	9,638	1,771	0.34	6,434	581	0.16	−22.76%
US	Mean	8,984	1,999	0.41	6,608	752	0.22	0.00%
	CoV	15.19%	24.51%	25.31%	36.22%	21.93%	46.79%	

efficiency and complexity are low. Philadelphia demonstrates the lowest construction efficiency at 22.76% below the national average.

Figure 8.4 shows the distribution of project types, with a dominant 68% in this case designated as primarily residential use. Mean citiBLOC cost/m^2 is shown in brackets.

The rate of increase over the past decade is +6.14% per annum (i.e. 33.236 / 541.23 × 100). Allowing for the growth in base costs of +3.20% per annum (i.e. 17.344 / 541.23 × 100), real construction efficiency is computed at +2.94% per annum. This is illustrated in Figure 8.5 using the dataset of 251 projects assembled according to their year of completion. The key findings of Chiang *et al.* (2012) are once again reflected in the United States.

Discussion

This data forms the basis for a useful comparison between Australian and American construction performance. Project cost data can be readily combined across cities and countries through the adoption of citiBLOC as an international locality index. In this research, 337 projects are included in the analysis (67% or two thirds of the projects that meet the set criterion of high-rise buildings completed between 2003 and 2012).

The citiBLOC data is effectively a construction-based PPP index. The standard basket is priced in local currency (i.e. AUD or USD) each year for each location. It enables costs to be compared among locations, including across national borders, without reference to a currency exchange rate. The notion of a city-based international locality index was applied in Langston and Best (2001; 2005), where they used the price of a Big Mac hamburger,

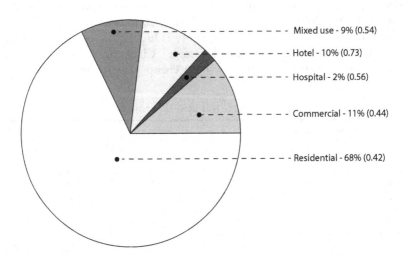

Figure 8.4 Project types (selected US projects)

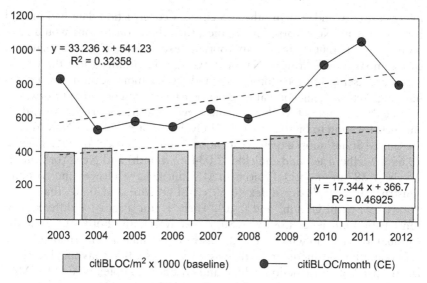

Figure 8.5 Construction efficiency over time (selected American projects)

sourced from *The Economist's Big Mac Index*, as a PPP index to adjust construction cost data. Interestingly, today the average price of a Big Mac in Australia is AUD$4.80 and in the United States is USD$4.20, while the mean citiBLOC index in Australia is AUD$10,000 and in United States is USD$8,984 – the ratio is nearly identical. But the Big Mac method did not work well in a number of developing countries where McDonald's hamburgers were more of a Western luxury item. The citiBLOC basket is more likely to be representative of global construction prices and is reasonably easy to calculate once per year.

Currency exchange rates rise and fall over time for a range of reasons, many of which have nothing at all to do with purchasing power. It is likely that the relative price of a citiBLOC in Australia and the United States has not changed much this century. The currency exchange rate, however, did change dramatically from 1 AUD = 0.5 USD in 2001 to 1 AUD = 1.08 USD in 2012, so conclusions about performance in the past based on exchange rates are quite misleading if quoted today. This is part of the problem when reviewing earlier research.

BCA (2012) compared the performance of large infrastructure projects in Australia and the United States and concluded that the former was uncompetitive. Included in their report were data on cost/m² for airports, schools, shopping malls and hospitals in both countries obtained from a well-known published cost guide. Apart from the obvious problems of using currency exchange rates and arguably not comparing 'like with like', as pointed out by Best (2012), they selected the US Gulf states as the comparative context to Australia. It can be seen from Table 8.4 earlier that Houston (Texas) has a much lower citiBLOC index than other US cities. If construction data had instead been used from New York, for instance, then their conclusions would have been quite different. Using an appropriate 'exchange rate' for international cost comparisons is critical. National averages in countries like the United States are useful, but location-specific indices are more accurate for benchmarking. Yet assessing comparative performance is still not straightforward.

Take the example of Melbourne and New York. Melbourne is Australia's cheapest location to build with a citiBLOC index of 9,754, and New York is the United States' most expensive location to build with a citiBLOC index of 10,693. Melbourne builds quickly (2,034 m²/month) and New York builds slowly (1,287 m²/month). Projects in Melbourne have a lower unit cost and construction complexity index (0.30 citiBLOCs/m² and 0.10) than New York (0.57 citiBLOCs/m² and 0.40). There is not a massive difference in construction efficiency (587 vs. 680, respectively), and both are below the national average. So which city demonstrates the higher performance?

PI might be considered to be the best ratio to use. But it favours locations where speed of construction is high and cost and complexity are low. Melbourne projects show such attributes (PI = 7,926). New York projects have the opposite attributes (PI = 2,963). CE gives a more balanced comparison. Despite variations in the complexity index, the ratio of cost over time is less sensitive. Based on the projects studied, New York is slightly ahead of Melbourne when assessing overall industry efficiency and on a par with both Sydney and Brisbane/GC.

CE provides a mechanism to compare the performance of both Australian and American construction industries based on microeconomic (i.e. project-level) data. It is concluded that, based on data from the largest five cities in each country, efficiency on site is improving in both countries. The growth in baseline cost/m² suggests a possible rise in project complexity over time. While the trend in efficiency improvement is similar, there is evidence that base costs in Australia have outstripped the United States, meaning that 'real' construction efficiency in Australia is relatively less. If Australia held an advantage in the past, then it seems that advantage might be disappearing. The United States is outperforming Australia in terms of construction efficiency by 1.10% per annum.

Differences in quality, such as fit-out or shell, are effectively eliminated, assuming cost and time vary in proportion to each other. This is probably an over-simplification, particularly as different building types are being mixed together. Nevertheless, high-rise construction is common to all projects in this study and is arguably a more dominant attribute than functional purpose. Figure 8.6 shows the correlation between cost and building height across the entire dataset. The higher the building, as you would expect, the higher is the cost to construct it, but the moderate value of R^2 only explains 33% of the relationship between the variables.

From the assembled database, the relationship among key variables like time, cost and quality can be explored in detail. The large number of projects

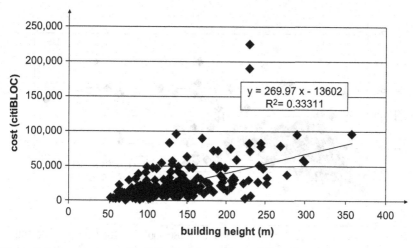

Figure 8.6 Comparison of building height and cost (all projects)

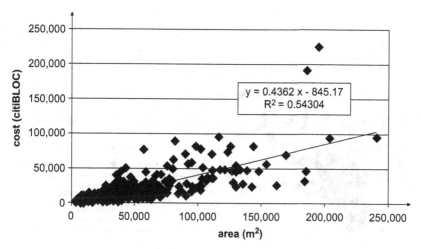

Figure 8.7 Comparison of area and cost (all projects)

reduces the influence of outliers, such as the two projects in Figure 8.6; however, outliers are of great interest, as they may represent examples of best (or worst) practice.

Figure 8.7 shows the relationship between area and cost. A robust value for R^2 suggests that gross floor area is a better predictor of construction cost than building height, explaining more than 50% of the relationship between the variables.

Figure 8.8 compares floor area with time to construct, while Figure 8.9 compares cost with time to construct. In both cases, a similar

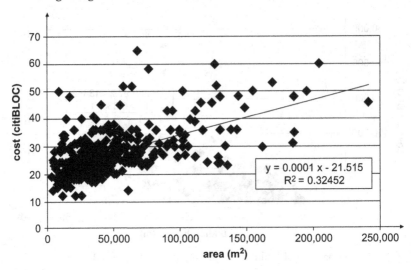

Figure 8.8 Comparison of area and time (all projects)

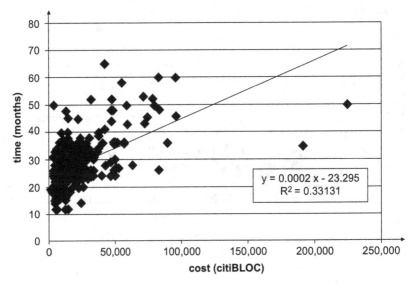

Figure 8.9 Comparison of cost and time (all projects)

result is found to Figure 8.6. Furthermore, time and building height (not shown) also share a modest R^2 of about 33%. It is concluded that while area is a reasonable predictor of cost, neither cost, area nor building height acts individually as a reliable predictor of time to construct for this building type. Previous research by Bromilow (1969) and Love *et al.* (2005) may not apply reliably or consistently for modern high-rise construction.

Comparing construction time between Australia and the United States, however, should consider the different industrial landscapes. Notionally, American construction workers have a 40-hour week, while Australian construction workers have a 38-hour week with 1 day in 20 being decreed a 'paid' rostered day off. Therefore, taking the 40-hour week as a base, adjusted Australian PI and CE scores could be as much as 5% higher than currently shown. This has no effect on their rate of change over time.

Conclusion

The main conclusion to be drawn from this research is that the efficiency of the Australian and US construction industries has increased at a similar rate over the past decade, although baseline costs have risen faster in Australia. Real construction efficiency, therefore, may be rising at about 1% per annum faster in the United States, and evidence exists to suggest that its top projects outperform anything found in Australia. But the key question that remains unanswered is why? Understanding the factors that drive efficiency in each country will hopefully shed light on what improvements are possible. These factors may be technical, political or contextual.

This research advances the notion that construction efficiency at a project level can be aggregated to determine construction efficiency of a contractor, a city or a nation. CE is computed as cost over time and intuitively assumes that complexity (quality and buildability) largely cancels out as both cost and time increase as complexity rises. Some evidence of this effect can be seen in the comparison between New York and Melbourne. However, the data for CC shows no correlation with time to construct. Why this should be so for high-rise construction remains a matter requiring further investigation.

A limitation of this study is that much of the key data concerning project cost, floor area and time to construct is yet to be validated. The enormity of the task means that only a small sample of projects in each country can be scrutinized. Cost data will be refined to ensure that design fees, site works, demolition and fit-out costs are excluded and that the construction cost is reflective of the final reconciliation for the project after all variations. Area data will be consistently measured from a standard definition including proper allowance for unenclosed covered floor area like balconies. Time to construct will be calculated from commencement on site to handover, with deductions for closed sites due to bankruptcy, as happened during the recent global financial crisis, but still including time lost due to industrial disputation, accident investigation and bad weather. Decisions to work longer hours per week via overtime payments may increase production output (m²/month) but will also increase resource input (cost/m²), so PI is not likely to change significantly. However, excessive use of overtime will improve CE scores and may be one reason for differences in perceived efficiency between projects.

It might be tempting for some to conclude that construction efficiency in Australia is actually higher than the United States on the basis of performance in a particular year, such as 2012. The volatility of the time series, to some extent dictated by the number of projects completed in a given year, suggests that a long-term perspective should be taken and hence why it is appropriate to employ linear regression to compute the trend in real construction efficiency over a 10-year period.

The relationships among cost, time and building height indicate each is correlated to the other, explaining about one third of the relationship in each case. Floor area is by far the most robust predictor of cost and time, while complexity and time have no observed correlation. The CoV of citiBLOC cost/m² values between each building type, computed at about 50% in both countries, is offset by the 337 data points that provide a more robust correlation test.

Finally, this research demonstrates the application of citiBLOC as a construction-relevant PPP index. At a national level, with Australia set at a base of 1, the United States is 0.8984. But more importantly, citiBLOC provides different indices for different cities, enabling locational variations in construction materials, labour and plant to be properly considered. The method for computing international locality indices represents a major advance in future construction performance studies and is relatively easy and practical to compile on an annual basis.

References and Further Reading

Abdelsalam, H.M.E. and Gad, M.M. (2009) Cost of quality in Dubai: an analytical case study of residential construction projects. *International Journal of Project Management*, **27** (5), 501–511.

Ali, A.S. and Rahmat, I. (2010) The performance measurement of construction projects managed by ISO-certified contractors in Malaysia. *Journal of Retail and Leisure Property*, **9** (1), 25–35.

Allan, C., Dungan, A. and Peetz, D. (2010) 'Anomalies', damned 'anomalies' and statistics: construction industry productivity in Australia. *Journal of Industrial Relations*, **52** (1), 61–79.

BCA (2012) *Pipeline or Pipe Dream? Securing Australia's Investment Future* (Business Council of Australia). www.bca.com.au/Content/101987.aspx

Best, R. (2012) International comparisons of cost and productivity in construction: a bad example. *Australasian Journal of Construction Economics and Building*, **12** (3), 82–88.

Bromilow, F.J. (1969) Contract time performance: expectations and reality. *Building Forum*, **1** (3), 70–80.

Brown, A. and Adams, J. (2000) Measuring the effect of project management on construction outputs: a new approach. *International Journal of Project Management*, 18 (5), 327–335.

Chan, P. W. and Kaka, A. (2007) Productivity improvements: understand the workforce perceptions of productivity first. *Personnel Review*, 36 (4), 564–584.

Chang, L. M. (1991) Measuring construction productivity: a case study based on the application of information in CII Source Document No. 35 and Thomas and Sanders' manual for collecting productivity and related data. *Cost Engineering*, 33 (10), 19–25.

Chau, K. W. (1993) Estimating industry-level productivity trends in the building industry from building cost and price data. *Construction Management and Economics*, 11 (5), 370–383.

Chiang, Y. H., Li, J., Choi, T.N.Y. and Man, K. F. (2012) Comparing China Mainland and China Hong Kong contractors' productive efficiency: a DEA Malmquist Productivity Index approach. *Journal of Facilities Management*, 10 (3), 179–197.

Crawford, P. and Vogl, B. (2006) Measuring productivity in the construction industry. *Building Research and Information*, 34 (3), 208–219.

Doloi, H. (2007) Twinning motivation, productivity and management strategy in construction projects. *Engineering Management Journal*, 19 (3), 30–40.

Egan, J. (1998) *Rethinking construction* (London: Department of the Environment, Transport and the Regions).

Enshassi, A., Mohamed, S., Mayer, P. and Abed, K. (2007) Benchmarking masonry labor productivity. *International Journal of Productivity and Performance Management*, 56 (4), 358–368.

Hewage, K. N. and Ruwanpura, J. Y. (2006) Carpentry worker issues and efficiencies related to construction productivity in commercial construction projects in Alberta. *Canadian Journal of Civil Engineering*, 33 (8), 1075–1089.

Horta, I. M., Camanho, A. S. and Moreira da Costa, J. (2012) Performance assessment of construction companies: a study of factors promoting financial soundness and innovation in the industry. *International Journal of Production Economics*, 137 (1), 84–93.

Hsieh, H. Y. and Forster, J. (2006) Residential construction quality and production levels in Taiwan. *Engineering, Construction and Architectural Management*, 13 (5), 502–520.

Kaming, P. F., Holt, G. D., Kometa, S. T. and Olomolaiye, P. O. (1998) Severity diagnosis of productivity problems: a reliability analysis. *International Journal of Project Management*, 16 (2), 107–113.

Langston, C. and Best, R. (2001) An investigation into the construction performance of high-rise commercial office buildings worldwide based on productivity and resource consumption. *International Journal of Construction Management*, 1 (1), 57–76.

Langston, C. and Best, R. (2005) Using the Big Mac Index for comparing construction costs internationally. In: *Proceedings of QUT Research Week*. Brisbane, July.

Latham, M. (1994) *Constructing the team* (London: HMSO).

Liao, P. C., Thomas, S. R., O'Brien, W. J., Dai, J., Mulva, S. P. and Kim, I. (2012) Benchmarking project level engineering productivity. *Journal of Civil Engineering and Management*, 18 (2), 235–244.

Lin, M. C., Tserng, H. P., Ho, S. P. and Young, D. L. (2011) Developing a construction-duration model based on a historical dataset for building project. *Journal of Civil Engineering and Management*, 17 (4), 529–539.

Loosemore, M. (2003) Impediments to reform in the Australian building and construction industry. *Australasian Journal of Construction Economics and Building*, 3 (2), 1–8.

Love, P.E.D., Tse, R.Y.C. and Edwards, D.J. (2005) Time-cost relationships in Australian building construction projects. *Journal of Construction Engineering and Management*, 131 (2), 187–194.

Low, S.P. (2001) Quantifying the relationships between buildability, structural quality and productivity in construction. *Structural Survey*, 19 (2), 106–112.

Meng, X. (2012) The effect of relationship management on project performance in construction. *International Journal of Project Management*, 30 (2), 188–198.

Mohamed, S. (1996) Benchmarking and improving construction productivity. *Benchmarking for Quality Management and Technology*, 3 (3), 50–57.

Mohamed, S. and Srinavin, K. (2002) Thermal environment effects on construction workers' productivity. *Work Study*, 51 (6/7), 297–302.

Motwani, J., Kumar, A. and Novakoski, M. (1995) Measuring construction productivity: a practical approach. *Work Study*, 44 (8), 18–20.

Munro-Faure, L. and Malcolm, M. (1992) *Implementing total quality management* (London: Pitman).

Pieper, P. (1989) Why construction industry productivity is declining: comment. *The Review of Economics and Statistics*, 71 (3), 543–546.

Rankin, J., Fayek, A.R., Meade, G., Haas, C. and Manseau, A. (2008) Initial metrics and pilot program results for measuring the performance of the Canadian construction industry. *Canadian Journal of Civil Engineering*, 35 (9), 894–907.

RBA (2012) *Inflation Calculator*. Reserve Bank of Australia. www.rba.gov.au/calculator/annualDecimal.html

Rosenfeld, Y. (2009) Cost of quality versus cost of non-quality in construction: the crucial balance. *Construction Management and Economics*, 27 (2), 107–117.

Sahay, B.S. (2005) Multi-factor productivity measurement model for service organisation. *International Journal of Productivity and Performance Management*, 54 (1), 7–22.

Shouke, C., Zhuobin, W. and Jie, L. (2010) Comprehensive evaluation for construction performance in concurrent engineering environment. *International Journal of Project Management*, 28 (7), 708–718.

Skyscraper (2012) *Skyscraper: Global Cities & Buildings Database*. http://skyscraperpage.com/cities/

Stewart, R.A. and Spencer, C.A. (2006) Six-sigma as a strategy for process improvement on construction projects: a case study. *Construction Management and Economics*, 24 (4), 339–348.

Tchidi, M.F., He, Z. and Li, Y.B. (2012) Process and quality improvement using six sigma in construction industry. *Journal of Civil Engineering and Management*, 18 (2), 158–172.

Walker, D.H.T. (1995) An investigation into construction time performance. *Construction Management and Economics*, 13 (3), 263–274.

Wegelius-Lehtonen, T. (2001) Performance measurement in construction logistics. *International Journal of Production Economics*, 69 (1), 107–116.

Willis, C.J. and Rankin, J.H. (2012) The construction industry macro maturity model (CIM3): theoretical underpinnings. *International Journal of Productivity and Performance Management*, 61 (4), 382–402.

Yang, H., Yeung, J.F.Y., Chan, A.P.C., Chiang, Y.H. and Chan, D.W.M. (2010) A critical review of performance measurement in construction. *Journal of Facilities Management*, 8 (4), 269–284.

Yang, L.R. (2009) Impacts of automation technology on quality of project deliverables in the Taiwanese construction industry. *Canadian Journal of Civil Engineering*, 36 (3), 402–421.

Yates, J.K. and Guhathakurta, S. (1993) International labor productivity. *Cost Engineering*, 35 (1), 15–25.

Editorial comment

When any attempt is made to compare the workings of the construction industries of different countries, more or less regardless of the precise method used, it is necessary to find a suitable way to make the construction costs that are expressed in various national currencies directly comparable. This is true whether the costs are at the level of basic resources such as concrete and steel or at a macro level where the value of construction is recorded for industry sectors or national industries.

In the previous chapter, the author introduced the idea of the citiBLOC, a relatively small basket of inputs to construction comprising five basic materials, four classes of labour and one item of equipment. The citiBLOC can be used as a unit of construction cost that is independent of exchange rates. In this chapter, the basic concept of the citiBLOC is expanded and tested using data collected for five major cities in four countries. The results are tested against those obtained using several other approaches. There is also further analysis and discussion of the theory behind purchasing power parities with particular reference to the context of building and construction.

It is suggested that the citiBLOC has a number of positive attributes that make it a preferred approach. Two key attributes are that it provides indices that are city based and that the data collection that is required to produce them is relatively simple. The localized nature of the citiBLOC index enables comparisons between cities and/or regions within countries; this is significant in countries such as the US where there are marked differences in construction costs between cities and national averages are problematic. The four-way comparison presented is based on data for the citiBLOC that is either available in commercial price books (or online versions of such books) and/or from industry people in-country. The small number of items in the citiBLOC basket means that collecting input costs in places where price books are not readily accessible is less of a challenge than it may be when prices are sought for a more extensive basket of items.

The analysis presented here supports the notion that a basket of inputs does not need to be large in order to provide a robust mechanism for construction cost conversions and this is a significant outcome.

9 Refining the citiBLOC index

Craig Langston

Introduction

The previous chapter investigated the performance of selected high-rise construction projects completed between 2003 and 2012 in the five largest cities in Australia and the United States. Part of this research required the conversion of cost data into a comparable form. A standard basket of construction material, labour and plant, priced in each city, was used for this purpose. The value of the standard basket (defined as equal to one 'citiBLOC') became the unit of cost comparison. For example, if a building's construction cost was AUD$10 million and the citiBLOC index for its location was AUD$10,000, then the 'cost' of the project would be computed as 1,000 citiBLOCs. Similarly, if a building was USD$16 million and the citiBLOC index was USD$8,000, then the 'cost' would be 2,000 citiBLOCs. The number of citiBLOCs can be used to translate local currency cost data into a comparable form. If the building floor areas were the same, then the US example would have a cost/m^2 rate that was twice as expensive as the Australian example.

Cost conversion is critical to international performance comparisons. A reliable method for doing this is necessary. Purchasing power parity (PPP) is generally accepted as the appropriate philosophy, but there are a number of detailed approaches for determining indices. Which one should be used? How can we be sure that cost conversion using currency exchange rates is unreliable? How should construction-related PPPs be produced for different countries and/or cities? What other issues should be considered?

The aim in this chapter is to answer these questions. Five approaches are discussed and explored for a representative range of countries. Australia and the United States are used as examples of small and large developed countries, respectively, whilst Malaysia and India are used as examples of small and large developing countries, respectively. Base data is largely determined from public-domain cost information available as part of the *Turner and Townsend International Construction Cost Survey 2012* (Turner and Townsend 2012) and citiBLOC calculations undertaken by the Centre for Comparative Construction Research (CCCR) at Bond University, Australia.

Background

Whenever the performance of the construction industry is called into question, the immediate reaction, and rightly so, is to attempt to benchmark performance against other countries. Performance is a complex issue, however, and includes multiple criteria such as cost, time and quality, to name but a few. While cost and time might appear reasonably straightforward, subjective issues such as quality are much more difficult to assess.

International cost comparison methodology is the focus of this chapter. Previous pricing studies have employed a range of methodologies, such as estimating the cost of identical standard projects (actual or hypothetical) or comparisons of functionally similar projects taking into account local practice or a combination of both. In any event, the question soon arises as to how to compare costs on an equal basis, since whenever different currencies are involved, the cost impacts cannot be immediately understood. Costs vary for a range of factors, not the least of which is time, but the issue of location is to be explored here and is central to the need to compare costs across national borders.

The exchange rate adopted to compare costs arising from projects in different locations is a critical factor for the usefulness of results that come from any international benchmarking study. Applying currency exchange rates is an obvious choice, but these change frequently and do not provide confidence that the relativity between construction industries in two different countries is actually being assessed. For example, the Asian economic crisis triggered in 1997 could be used to conclude that the dramatically lower cost of construction in some Asian countries, as calculated by falling exchange rates against their Western counterparts, was a result of increased competitiveness in-country. The reality was that the local industry had not changed, but the value assigned to projects that were under construction or previously completed had sharply declined.

The use of PPP as an alternative to traditional currency exchange rates is generally regarded as a superior approach (e.g. Rogoff 1996; Langston and Best 2005). PPP is an attempt to measure the economic well-being of people according to the country in which they reside. While not pretending to be an indicator of living standards, it does reflect the cost of living in-country and therefore forms a new baseline against which construction costs can be interpreted.

PPPs can be calculated at the value of a particular good or service or using a weighted basket of goods and services and can be expressed in relation to gross domestic product or income capacity. In fact, PPPs have been calculated using items that are available in most countries worldwide, such as via use of the Big Mac Index regularly compiled by *The Economist* magazine. There are grounds to suggest that an approach specific to construction goods and services would be preferable to one that is generic of entire economies (e.g. Walsh and Sawhney 2004).

PPPs are defined as exchange rates that replace traditional currency exchange rates by taking into account the differences in prices between countries (Pakko and Pollard 2003). They convert local costs into 'international dollars' compared to a nominated base country. The philosophy behind PPPs is the *Law of One Price* – namely that the cost of a good or service should be the same in different countries – else people would buy goods cheaper from one country and sell them at a profit in another.

Whether the *Law of One Price* holds for any particular item depends on the item meeting four basic criteria (UBS 2003:27). They are:

1   The item must be tradable.
2   There are no impediments to trade.
3   There are no transaction costs (such as transport) involved in trade of the item.
4   The item is perfectly homogeneous across all locations.

If all four criteria are met, then the price of the item should be the same in different places at the same time. In that case, the cost of an item in currency X should represent the same value as the cost of the same item in currency Y (Best 2008).

The United Nations–sponsored International Comparison Program (ICP) commenced in 1967 and now produces PPPs published by the World Bank Group for most countries on an approximate three-year cycle. These indices have been interpreted and extended to form the Penn World Table (PWT) produced by the University of Pennsylvania. The Eurostat-OECD joint program currently collects more detailed PPP data than the ICP, but for a much smaller set of countries. Indices for both ICP and Eurostat-OECD PPPs are expressed as a proportion of per-capita gross domestic product. The Union Bank of Switzerland (UBS) has also been producing PPP data since 1970, again approximately on a three-year cycle. They use a basket of goods and services and express their data in three forms (using a base for Switzerland, the United States or the Euro-zone, respectively). One criticism of these programs is the time delay between data collection and publication. Another criticism is the cost of the process.

Controversially, *The Economist* magazine has published an alternate PPP index based on the McDonald's Big Mac hamburger price since 1986 for a number of countries. Known as 'burgernomics' (Lan 2003), this approach has moved from a lighthearted look at fast food metrics to a quite seriously debated topic (e.g. Pakko and Pollard 2003, who found a correlation of 0.73 between the PWT and the Big Mac Index using 2000 data). The Big Mac Index has the advantage that input data is relatively easy to collect and therefore enables it to be up to date and city specific. *The Economist* now publishes its index several times each year. Cumby (1996) found that when the US dollar price of a Big Mac is high in a country, the relative local currency price of a Big Mac in that country is likely to fall during the following year. The index has been employed to identify currency over and under valuations, although this is not a recommended use.

The reliability of various methods is unknown, as there is no correct value that each can be compared against other than monetary exchange rates, which are volatile and subject to influence from a number of external sources. Pakko and Pollard (2003:22) concluded that *"it is interesting to find that the simple collection of items comprising the Big Mac sandwich does just as well (or just as poorly) at demonstrating the principles and pitfalls of PPP as do more sophisticated measures"*. Ong (2003) concurred. But over the last decade in particular, attention has now turned to developing indices that are industry focused, such as comparing construction-related costs independent of general economy activity (Meikle 1990; Walsh and Sawhney 2004).

Approach

Five methods shall be compared as part of this study. Each will be applied to the four selected countries to highlight differences due to country size and affluence level. The methods are described briefly below:

1 *Method 1*: a mix of various base prices for the supply of construction labour and material, weighted according to their cost impact on building sites.
2 *Method 2*: an unweighted mix of composite prices for the supply and installation of various construction components commonly found in international building projects.
3 *Method 3*: a basket of location-specific common work items for material supply, labour charge-out rates and plant hire, averaged across the five largest cities by population.
4 *Method 4*: the current price of a standard specification McDonald's Big Mac hamburger.
5 *Method 5*: currency exchange rates.

The methods are selected based on their ability to expose important issues related to international cost conversion. Other variations on these methods are also possible. A method based on pricing identical projects in different countries was not tested, as it was too time consuming and unlikely to lead to satisfactory outcomes given that individual items of work in some countries will not be representative of local practice.

Across all methods, one country is nominated as the base and the PPP indices for the other countries are computed as price relatives to this base. In this study, Australia is the selected base country on the basis that it has the lowest coefficient of variation (CoV) across its five largest cities, as determined by Method 3 (see Table 9.9). The United States had the largest national variation.

Costs/m^2 are collected for a range of building types and then adjusted by the derived PPPs. Rather than use one or more building types individually, a large range of building types is averaged to increase statistical reliability. The average cost/m^2 for building in each country is then divided by the country PPP and multiplied by the base PPP, and the results are compared. A summary of each method per country is used to determine which method produces the lowest CoV and therefore which method best reflects the *Law of One Price*. The same approach was previously applied in Langston and Best (2005).

All data used in this study is from Turner and Townsend (2012) unless otherwise noted. Several errors found in this data are noted and amended. It is acknowledged that accuracy of this data is assumed to be acceptable for the purposes of testing the relative performance of the different methods. Nevertheless, accuracy of raw data is a limitation of international cost comparisons.

Results

Method 1 (L+M)

Turner and Townsend (2012) provide 5 items of construction labour and 11 items of construction material for Australia, the United States, Malaysia and India. These are national prices and do not reflect variations between different locations. Table 9.1 shows the raw data obtained from this source.

Table 9.2 converts the raw data in Table 9.1 to price relatives using Australia as the base. Labour and material means are then combined to form a single index. The ratio is derived from the data used in Method 3. The percentage of labour is computed as the sum of the four labour items in the citiBLOC basket divided by the sum of all 10 items plus 50% of the plant item. Similarly, the percentage of material is computed as the sum of the five material items in the citiBLOC basket divided by the sum of all 10 items plus the remaining 50% of the plant item. As the quantities used in Method 3 were set to ensure evenly weighted basket items, the ratios for Australia reflect the 50:40:10 design of the citiBLOC basket itself.

The definition of 'mean' used in this method and others in this study is the arithmetic mean. It would be inappropriate to use geometric mean, as the data is not a compounding time series. Arithmetic mean is also simpler to calculate and more understandable in practice.

Table 9.1 Method 1 raw data (Turner and Townsend 2012)

	Australia (AUD)	United States (USD)	Malaysia (MYR)	India (INR)
LABOUR *(/hour):*				
Group 1 tradesman	68	75	20	56
Group 2 tradesman	57	65	15	50
Group 3 tradesman	55	57	20	38
General labourer	38	53	10	20
Site foreman	72	77	55	80
MATERIAL:				
Concrete 30 MPa (m³)	186	135	230	5,000
Reinforcement bar 16mm (tonne)	1,250	992	3,220	40,500
Concrete block 400 × 200 (per 1,000)	3,360	1,030	3,980	35,000
Standard brick (per 1,000)	541	350	400	6,250
Structural steel beams (tonne)	2,955	1,150	4,420	50,000
Glass pane 6mm (m²)	47	58	189	500
Softwood timber for framing 100 × 50mm (m)	29	7	34	275
Plasterboard 13mm (m²)	39	3	26	280
Emulsion paint (litre)	15	8	39	340
Copper pipe 15mm (m)	12	7	27	^ 435
Copper cable 3C+E 2.5mm PVC (m)	5	4	16	^^ 25
COMBINED MIX *(%):*				
Labour	44.62	53.81	16.55	8.14
Material	55.38	46.19	83.45	91.86
sum	100.00	100.00	100.00	100.00

^ error fixed: /m not /kg ^^ error fixed: /m not /100m

Method 2 (composite)

Turner and Townsend (2012) further provide 19 composite items of construction for Australia, the United States, Malaysia and India. Again, these are national prices and do not reflect variations between different locations. Table 9.3 shows the raw data obtained from this source.

Table 9.4 converts the raw data in Table 9.3 to price relatives using Australia as the base.

Table 9.2 Method 1 price relatives (Australia = base)

	Australia (AUD)	United States (USD)	Malaysia (MYR)	India (INR)
LABOUR (*/hour*):				
Group 1 tradesman	100.00	110.29	29.41	82.35
Group 2 tradesman	100.00	114.04	26.32	87.72
Group 3 tradesman	100.00	103.64	36.36	69.09
General labourer	100.00	139.47	26.32	52.63
Site foreman	100.00	106.94	76.39	111.11
mean	100.00	114.88	38.96	80.58
MATERIAL:				
Concrete 30 MPa (m³)	100.00	72.58	123.66	2,688.17
Reinforcement bar 16mm (tonne)	100.00	79.36	257.60	3,240.00
Concrete block 400 × 200 (per 1,000)	100.00	30.65	118.45	1,041.67
Standard brick (per 1,000)	100.00	64.70	73.94	1,155.27
Structural steel beams (tonne)	100.00	38.92	149.58	1,692.05
Glass pane 6mm (m²)	100.00	123.40	402.13	1,063.83
Softwood timber for framing 100 × 50mm (m)	100.00	24.14	117.24	948.28
Plasterboard 13mm (m²)	100.00	7.69	66.67	717.95
Emulsion paint (litre)	100.00	53.33	260.00	2,266.67
Copper pipe 15mm (m)	100.00	58.33	225.00	3,625.00
Copper cable 3C+E 2.5mm PVC (m)	100.00	80.00	320.00	490.00
mean	100.00	57.76	192.21	1,720.81
COMBINED MIX (*%*):				
Labour	44.62	61.82	6.45	6.56
Material	55.38	26.58	160.39	1,580.69
sum	100.00	88.40	166.84	1,587.25

Method 3 (citiBLOC)

Method 3 is the citiBLOC index applied in the previous chapter. The five cities forming the Australian average are Sydney, Melbourne, Brisbane (including Gold Coast), Perth and Adelaide (listed in decreasing order of population size). In the United States, the cities are New York, Los Angeles, Chicago, Houston and Philadelphia. In Malaysia, the cities comprise Kuala

Table 9.3 Method 2 raw data (Turner and Townsend 2012)

	Australia (AUD)	United States (USD)	Malaysia (MYR)	India (INR)
SUPPLY AND INSTALL:				
Excavate basement (m³)	29	10	28	750
Excavate footings (m)	71	8	33	475
Concrete in slab (m³)	256	171	339	6,500
Reinforcement in beams (tonne)	2,442	2,100	3,578	56,000
Formwork to soffit of slab (m²)	120	71	51	525
Blockwork in wall (m²)	131	75	68	1,075
Structural steel beams (tonne)	6,298	2,855	6,632	82,500
Precast concrete wall (m²)	299	120	n/a	8,250
Curtain wall glazing incl. support system (m²)	1,051	775	408	9,250
Plasterboard 13mm thick to stud walls (m²)	30	34	92	2,250
Single solid core door incl. frame/hardware (no)	765	2,100	816	27,500
Painting to walls primer + 2 coats (m²)	15	10	8	200
Ceramic tiling (m²)	109	90	112	2,800
Vinyl flooring to wet areas (m²)	71	64	100	2,100
Carpet medium tufted (m²)	80	44	92	2,350
Lighting installation (m²)	81	83	153	2,900
Copper pipe 15mm to wall (m)	60	32	80	^ 600
Fire sprinklers (m²)	46	30	102	825
Air conditioning including main plant (m²)	300	172	306	3,625

^ error fixed: /m not /kg

Lumpur, Johor Bahru, Ipoh, Kuching and Georgetown. Finally, the cities forming the India average are Mumbai, New Delhi, Bangalore, Kolkata and Chennai. The total value of the citiBLOC basket in each of these cities is shown in Tables 9.5 to 9.8. The proportional mix of material, labour and plant by city is also provided based on the quantity of each work item. For the Australian base, the total value of each item is approximately equal value. The national figure is assumed to be the average of prices from the five largest cities.

Australian data was extracted from Cordells online cost database. American data was extracted from RS Means online cost database. Malaysian

Table 9.4 Method 2 price relatives (Australia = base)

	Australia (AUD)	United States (USD)	Malaysia (MYR)	India (INR)
SUPPLY AND INSTALL:				
Excavate basement (m^3)	100.00	34.48	96.55	2,586.21
Excavate footings (m)	100.00	11.27	46.48	669.01
Concrete in slab (m^3)	100.00	66.80	132.42	2,539.06
Reinforcement in beams (tonne)	100.00	86.00	146.52	2,293.20
Formwork to soffit of slab (m^2)	100.00	59.17	42.50	437.50
Blockwork in wall (m^2)	100.00	57.25	51.91	820.61
Structural steel beams (tonne)	100.00	45.33	105.30	1,309.94
Precast concrete wall (m^2)	100.00	40.13	n/a	2,759.20
Curtain wall glazing incl. support system (m^2)	100.00	73.74	38.82	880.11
Plasterboard 13mm thick to stud walls (m^2)	100.00	113.33	306.67	7,500.00
Single solid core door incl. frame/hardware (no)	100.00	274.51	106.67	3,594.77
Painting to walls primer + 2 coats (m^2)	100.00	66.67	53.33	1,333.33
Ceramic tiling (m^2)	100.00	82.57	102.75	2,568.81
Vinyl flooring to wet areas (m^2)	100.00	90.14	140.85	2,957.75
Carpet medium tufted (m^2)	100.00	55.00	115.00	2,937.50
Lighting installation (m^2)	100.00	102.47	188.89	3,580.25
Copper pipe 15mm to wall (m)	100.00	53.33	133.33	1,000.00
Fire sprinklers (m^2)	100.00	65.22	221.74	1,793.48
Air conditioning including main plant (m^2)	100.00	57.33	102.00	1,208.33
mean	100.00	75.51	118.43	2,251.00

and Indian data was provided by personal contacts in-country. Table 9.9 converts the raw data in Tables 9.5 to 9.8 to price relatives, again using Australia as the base.

Method 4 (Big Mac Index)

The price of an international standard commodity can be used as a handy PPP since, theoretically, it is of equal value in all countries. Yet this is unlikely for a number of reasons, including distortions to free trade, different market

Table 9.5 Method 3 raw data (Australia)

	citiBLOC (AUD)					
	Sydney	Melbourne	Brisbane	Perth	Adelaide	(National)
MATERIAL:						
32 MPa ready-mixed concrete (45m³)	11,144	8,764	9,907	10,559	10,109	10,097
Steel in 250 × 25.7kg/m 'I' beam (6.8t)	9,350	8,500	13,430	8,813	10,846	10,188
10mm clear tempered glass (44m²)	10,472	10,472	9,724	10,472	10,472	10,322
13mm-thick gypsum plasterboard (1300m²)	10,140	10,010	10,530	10,790	9,100	10,114
100 × 50mm sawn softwood stud (2750m)	9,873	9,873	9,873	9,873	9,873	9,873
LABOUR:						
Electrician (150hrs)	9,000	10,950	10,200	9,825	8,850	9,945
Carpenter (185hrs)	10,915	10,730	9,435	9,528	9,158	9,953
Painter (200hrs)	10,400	10,600	9,600	9,100	9,800	9,900
Unskilled labour (275hrs)	10,863	10,038	10,725	9,213	9,350	10,038
PLANT:						
50 t mobile crane (5days)	10,200	7,600	9,520	10,200	10,320	9,568
mean	10,326	9,754	10,294	9,837	9,778	10,000
PROPORTIONAL MIX (%):						
Material	49.37	48.82	51.93	51.34	51.49	55.38
Labour	40.75	43.39	38.82	38.29	37.96	44.62
Plant	9.88	7.79	9.25	10.37	10.55	incl.

contexts and demand, and local availability of key resources. A McDonald's Big Mac is an example of a standard commodity. This product is used here to compare against the industry-specific PPPs. The price of a Big Mac in each country for 2012 is provided in Table 9.10. Data was sourced from *The Economist* magazine and reflects the mean national price for the year.

Table 9.6 Method 3 raw data (United States)

	citiBLOC (USD)					
	New York	Los Angeles	Chicago	Houston	Philadelphia	(National)
MATERIAL:						
32 MPa ready-mixed concrete (45m³)	5,887	5,471	6,055	5,061	5,602	5,615
Steel in 250 × 25.7kg/m 'I' beam (6.8t)	16,062	14,049	12,281	13,750	13,634	13,955
10mm clear tempered glass (44m²)	8,557	8,228	7,891	8,648	8,351	8,335
13mm-thick gypsum plasterboard (1300m²)	3,796	3,445	3,276	3,055	3,341	3,383
100 × 50mm sawn softwood stud (2750m)	3,630	3,218	3,108	2,970	3,273	3,240
LABOUR:						
Electrician (150hrs)	14,066	9,536	10,334	5,423	16,541	11,180
Carpenter (185hrs)	15,314	9,984	13,085	5,437	11,453	11,055
Painter (200hrs)	11,942	8,386	12,300	6,562	11,286	10,095
Unskilled labour (275hrs)	12,279	10,673	9,336	8,542	9,897	10,145
PLANT:						
50 t mobile crane (5days)	15,400	12,600	12,600	10,600	13,000	12,840
mean	10,693	8,559	9,026	7,005	9,638	8,984
PROPORTIONAL MIX (%):						
Material	35.47	40.20	36.13	47.80	35.49	46.19
Labour	50.13	45.07	49.91	37.07	51.03	53.81
Plant	14.40	14.73	13.96	15.13	13.48	incl.

Table 9.7 Method 3 raw data (Malaysia)

	citiBLOC (MYR)					
	Kuala Lumpur	Johor Bahru	Ipoh	Kuching	Georgetown	(National)
MATERIAL:						
32 MPa ready-mixed concrete (45m³)	9,900	9,450	10,350	11,250	9,900	10,170
Steel in 250 × 25.7kg/m 'I' beam (6.8t)	20,264	17,000	21,080	23,120	20,264	20,346
10mm clear tempered glass (44m²)	10,560	9,680	11,000	11,440	9,240	10,384
13mm-thick gypsum plasterboard (1300m²)	39,000	33,800	36,400	46,800	36,400	38,480
100 × 50mm sawn softwood stud (2750m)	19,250	19,250	22,000	19,250	19,250	19,800
LABOUR:						
Electrician (150hrs)	3,750	3,000	3,750	4,200	3,750	3,690
Carpenter (185hrs)	3,700	2,775	3,700	4,255	3,700	3,626
Painter (200hrs)	5,000	4,000	5,000	5,600	5,000	4,920
Unskilled labour (275hrs)	4,125	2,750	4,125	4,950	4,125	4,015
PLANT:						
50 t mobile crane (5days)	9,000	7,500	9,000	9,000	9,000	8,700
mean	12,455	10,921	12,641	13,987	12,063	12,413
PROPORTIONAL MIX (%):						
Material	79.47	81.66	79.77	79.98	78.80	83.45
Labour	13.31	11.47	13.11	13.59	13.74	16.55
Plant	7.22	6.87	7.12	6.43	7.46	incl.

Method 5 (Currency exchange rates)

Currency exchange rates are not PPPs. However, this method is included to benchmark results against a popular, albeit inappropriate, strategy for international cost comparisons. The exchange rates at 30 June 2012, sourced online (www.oanda.com/currency/classic-converter) are shown in Table 9.11.

Table 9.8 Method 3 raw data (India)

	citiBLOC (INR)					
	Mumbai	New Delhi	Bangalore	Kolkata	Chennai	(National)
MATERIAL:						
32 MPa ready-mixed concrete (45m³)	214,875	163,305	182,655	243,360	206,280	202,095
Steel in 250 × 25.7kg/m 'I' beam (6.8t)	310,080	323,000	335,920	291,992	384,200	329,038
10mm clear tempered glass (44m²)	80,520	61,600	48,136	81,400	106,524	75,636
13mm-thick gypsum plasterboard (1300m²)	244,400	210,500	188,500	292,500	214,500	228,280
100 × 50mm sawn softwood stud (2750m)	288,750	288,750	390,500	316,250	269,500	310,750
LABOUR:						
Electrician (150hrs)	6,563	5,363	8,438	6,563	9,375	7,260
Carpenter (185hrs)	7,631	6,938	10,406	8,094	11,563	8,926
Painter (200hrs)	8,750	7,500	11,250	8,750	12,500	9,750
Unskilled labour (275hrs)	8,594	6,875	10,313	7,563	12,031	9,075
PLANT:						
50 t mobile crane (5days)	150,000	140,000	150,000	140,000	150,000	146,000
citiBLOC	132,016	120,483	133,612	139,647	137,647	132,681
PROPORTIONAL MIX (%):						
Material	86.25	86.17	85.75	87.76	85.80	91.86
Labour	2.39	2.21	3.02	2.22	3.30	8.14
Plant	11.36	11.62	11.23	10.02	10.90	incl.

Comparison of methods

Turner and Townsend (2012) provide construction costs/m² for nine different building types in each of the four selected countries. These are listed in Table 9.12. Rather than compute each building type separately, the mean of all building types was adopted to provide a more stable cost benchmark.

Table 9.9 Method 3 price relatives (Australia = base)

	Australia (AUD)	United States (USD)	Malaysia (MYR)	India (INR)
LOCATION:				
City 1 (largest)	103.26	106.93	124.55	1,320.16
City 2	97.54	85.59	109.21	1,204.83
City 3	102.94	90.26	126.41	1,336.12
City 4	98.37	70.05	139.87	1,396.47
City 5	97.88	96.38	120.63	1,376.47
mean	100.00	89.84	124.13	1,326.81
CoV	2.85%	15.19%	8.90%	5.63%

Table 9.10 Method 4 raw data

	Australia (AUD)	United States (USD)	Malaysia (MYR)	India (INR)
INTERNATIONAL STANDARD PRODUCT:				
Big Mac hamburger	4.80	4.20	7.35	84.00

Table 9.11 Method 5 raw data

	Australia (AUD)	United States (USD)	Malaysia (MYR)	India (INR)
30 JUNE 2012:				
Currency exchange rate (compared to USD)	0.9844	1.0000	3.1989	56.225

Equivalent costs/m^2 are shown in Table 9.13. The CoV can be used to rank the reliability of each method. The lower the CoV, the more the method reflects the *Law of One Price* philosophy.

Method 1 has an issue in determining the proportional mix of labour and material, which varies between low-labour and high-labour countries and is difficult to quantify. Method 2 overcomes that problem but also embeds issues of productivity rather than being merely a cost conversion tool (i.e. higher comparative costs may indicate poor productivity, complex design or expensive resource inputs). Method 3 solves both of those shortcomings and has the added advantage of being an effective locality index. Method 4 delivers a quick and up-to-date general PPP index, but it may

Table 9.12 Cost/m² for various building types (Turner and Townsend 2012)

	Australia (AUD)	United States (USD)	Malaysia (MYR)	India (INR)
BUILDING TYPE:				
Residential	1,990	1,649	1,700	29,250
Commercial	2,660	2,182	3,167	34,333
Warehouses	963	1,262	2,067	33,333
Retail	2,727	1,283	3,333	27,167
Hotels	3,702	2,158	5,767	53,667
Hospitals	3,806	2,898	2,300	29,000
Schools	2,500	1,885	1,650	28,000
Car parks	997	1,008	1,450	31,000
Airports	6,565	3,550	5,000	65,000
mean	2,879	1,986	2,937	36,750

Table 9.13 Cost/m² price relatives (Australia = base)

	Australia	United States	Malaysia	India	CoV (%)
METHOD:					
Method 1 (L+M)	2,879	2,247	1,760	2,315	19.91%
Method 2 (composite)	2,879	2,630	2,480	1,633	22.48%
Method 3 (citiBLOC)	2,879	2,211	2,366	2,770	12.48%
Method 4 (Big Mac Index)	2,879	2,270	1,918	2,100	18.19%
Method 5 (currency exchange rates)	2,879	1,955	904	643	64.35%

not always be applicable or defendable for construction work (particularly for developing countries). Method 5 is problematic, as currency conversion can change due to global macroeconomic factors irrespective of local purchasing power.

Method 3 has the lowest coefficient of variation (12.48%) and is recommended as more likely to reflect construction-related PPP. Method 5 has the highest coefficient of variation (64.35%) and should not be used. Method 4 appears to produce results slightly superior to Methods 1 and 2 in the countries studied. Figure 9.1 shows the profile of the various methods.

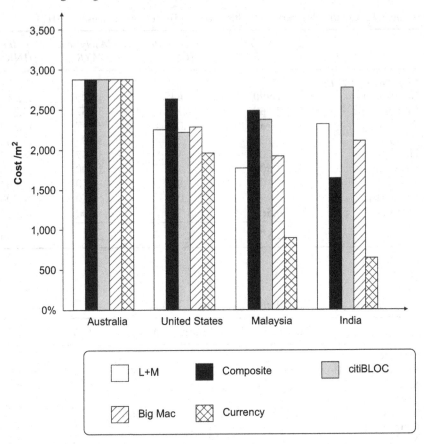

Figure 9.1 Comparison of construction price relatives

The profile of methods is the same regardless of the accuracy of the average costs/m² in each country. Further testing is nevertheless warranted, including a broader range of countries studied.

Discussion

In this chapter, citiBLOC indices for cities were calculated as an evenly weighted basket of construction material, labour and plant. The quantities for each item were chosen to enable the items to have equal influence. This is certainly the case for Australia, but other countries show that the balance is lost. This appears to be a disadvantage of Method 3 and may introduce bias in the results. However, the quantities chosen are irrelevant to the outcomes.

Table 9.14 Unit rates for citiBLOC for Australia

	AUD/unit					
	Sydney	*Melbourne*	*Brisbane*	*Perth*	*Adelaide*	*(National)*
MATERIAL:						
32 MPa ready-mixed concrete (m³)	247.65	194.75	220.15	234.65	224.65	224.37
Steel in 250 × 25.7kg/m 'I' beam (t)	1,375.00	1,250.00	1,975.00	1,296.00	1,595.00	1,498.20
10mm clear tempered glass (m²)	238.00	238.00	221.00	238.00	238.00	234.60
13mm-thick gypsum plasterboard (m²)	7.80	7.70	8.10	8.30	7.00	7.78
100 × 50mm sawn softwood stud (m)	3.59	3.59	3.59	3.59	3.59	3.59
LABOUR:						
Electrician (hr)	66.00	73.00	68.00	65.50	59.00	66.30
Carpenter (hr)	59.00	58.00	51.00	51.50	49.50	53.80
Painter (hr)	52.00	53.00	48.00	45.40	49.00	49.50
Unskilled labour (hr)	39.50	36.50	39.00	33.50	34.00	36.50
PLANT:						
50 t mobile crane (day)	2,040.00	1,520.00	1,904.00	2,040.00	2,064.00	1,913.60

The adoption of price relatives means that the price of a particular item is compared to the base, and the quantity multiplier cancels out. Table 9.14 shows the calculation for Australia and the base for each of the 10 items in the citiBLOC basket against which unit rates in other countries are compared.

The unit rates for each country are expressed as a ratio of the Australian average and converted to a base of 100. The procedure is illustrated in Table 9.15 using India as an example. The difference between 1,324.63 (Table 9.15) and the previous figure for India of 1,326.81 (see Table 9.9) is merely rounding error. Therefore the use of quantities as part of Method 3 does not introduce bias and is really only done to ensure that prices are reflective of a given scope of work. Note that the proportional mix of material, labour and plant (see Table 9.8) cannot be calculated from unit rate data.

Table 9.15 Example price relatives for India (based on unit rates)

	INR/unit					
	Mumbai	*New Delhi*	*Bangalore*	*Kolkata*	*Chennai*	*(National)*
MATERIAL:						
32 MPa ready-mixed concrete (m³)	2,128.18	1,617.42	1,809.07	2,410.30	2,043.05	2,001.60
Steel in 250 × 25.7kg/m 'I' beam (t)	3,043.65	3,170.47	3,297.29	2,866.11	3,771.19	3,229.74
10mm clear tempered glass (m²)	780.05	596.76	466.33	788.58	1,031.97	732.74
13mm-thick gypsum plasterboard (m²)	2,416.45	1,992.29	1,863.75	2,892.03	2,120.82	2,257.07
100 × 50mm sawn softwood stud (m)	2,924.79	2,924.79	3,955.43	3,203.34	2,729.81	3,147.63
LABOUR:						
Electrician (hr)	65.99	53.92	84.84	65.99	94.27	73.00
Carpenter (hr)	76.67	69.70	104.55	81.32	116.17	89.68
Painter (hr)	88.38	75.76	113.64	88.38	126.26	98.48
Unskilled labour (hr)	85.62	68.49	102.74	75.34	119.86	90.41
PLANT:						
50 t mobile crane (day)	1,567.73	1,463.21	1,567.73	1,463.21	1,567.73	1,525.92
mean	1,317.75	1,203.28	1,336.54	1,393.46	1,372.11	1,324.63

Conclusion

The introduction to this chapter raised a number of questions. Which PPP method should be used? How can we be sure that cost conversion using currency exchange rates is unreliable? How should construction-related PPPs be produced for different countries and/or cities? What other issues should be considered?

The answers to these questions are now clear. The recommended construction-related PPP method is identified as Method 3. Known as citi-BLOC, it uses a basket of 10 common construction items (5 material, 4 labour and 1 plant) priced per city, with the country average being computed from the five largest cities (where applicable). This method has the lowest CoV of the five methods tested and therefore is more likely to reflect the *Law of One Price* underpinning the PPP philosophy. It is also shown that current exchange rates have the highest CoV, which demonstrates why they should not be used for international cost comparisons. The procedure for calculating citiBLOCs for different locations is outlined, including proof that the quantities assigned to the basket make no difference to the outcome and are used purely to signify pricing context.

The citiBLOC index is refined in this chapter. One citiBLOC is defined as the equivalent of AUD$10,000 in 2012 measured across the five largest cities. It needs to be computed each year. This value typically rises each year due to the effects of inflation. It can be used as a benchmark of construction cost for any location, whether within a particular country (i.e. regional comparisons) or across different countries (i.e. international comparisons).

References and Further Reading

Best, R. (2008) *Development and testing of a purchasing power parity method for comparing construction costs internationally.* University of Technology, Sydney: Unpublished PhD thesis.

Croce, N., Green, R., Mills, B. and Toner, P. (1999) *Constructing the future: a study of major building construction in Australia.* University of Newcastle, Australia: Employment Studies Centre.

Cumby, R. E. (1996) *Forecasting exchange rates and relative prices with the hamburger standard: is what you want what you get with McParity?* NBER Working paper 5675 (Cambridge: National Bureau of Economic Research).

Lan, Y. (2003) *The long-term behaviour of exchange rates (part IV): Big Macs and the evolution of exchange rates.* Discussion Paper 03.08 University of Western Australia: Business School.

Langston, C. and Best, R. (2000) *International construction study* (Canberra: Department of Industry Science and Resources).

Langston, C. and Best, R. (2001) An investigation into the construction performance of highrise commercial office buildings based on productivity and resource consumption. *International Journal of Construction Management,* 1 (1), 57–76.

Langston, C. and Best, R. (2005) Using the Big Mac Index for comparing construction costs internationally. In: *Proceedings of QUT Research Week.* Queensland University of Technology, Brisbane, July.

Langston, C. and de Valence, G. (1999) *International cost of construction study – stage 2: evaluation and analysis* (Canberra: Department of Industry Science and Resources).

Meikle, J. (1990) International comparisons of construction costs and prices. *Habitat International,* 14 (2/3), 185.

Ong, L. L. (2003) *The Big Mac index: applications of purchasing power parity* (New York: Palgrave Macmillan).

Pakko, M. R. and Pollard, P. S. (2003) Burgernomics: a Big Mac™ guide to purchasing power parity. *Federal Reserve Bank of St. Louis Review,* 85 (6), 9–28.

Rogoff, K. (1996) The purchasing power puzzle. *Journal of Economic Literature*, **XXXIV** (2), 647–668.

Turner & Townsend (2012) *International construction cost survey 2012:* www.turnerandtownsend.com/construction-cost-2012/_16803.html

UBS (2003) *Prices and earnings: a comparison of purchasing power parity around the globe* (2003 Edition) (Switzerland: Union Bank of Switzerland).

Walsh, K. and Sawhney, A. (2004) *International comparison of cost for the construction sector: an implementation framework for the basket of construction components approach*. Report submitted to the African Development Bank and the World Bank Group.

Editorial comment

Productivity is often defined simply as 'output over input' or, more specifically, as the ratio of an amount of output and the resources employed to produce that amount of output. In some circumstances it is relatively easily measured: for example, site activities such as the laying of bricks can be measured in terms of the number of units laid in a day by a given number of workers. Even such a simple measure, however, has limitations, as the speed or efficiency of the bricklayers may be affected by things such as working at heights, climate or weather conditions, the size and type of the units being laid and the required and/or achieved quality of work.

At least in that example, the volume of output and the key resource (time, expressed in hours or days) are easily measured, but measuring construction output at higher levels is not so straightforward. In this chapter the authors review several decades of research aimed at the measurement of construction productivity above the level of basic on-site activities. The review highlights several things, not the least of which is that there have been many attempts to measure construction industry productivity over the past 50 years, and these attempts have been based on a wide variety of methodologies.

In many cases the drivers for these productivity measures have been claims that productivity growth in construction has stalled and/or lagged behind that of other industries such as manufacturing and agriculture, or even that construction productivity has declined over time. Some of these claims may be driven by vested interests such as industry bodies looking for ways to exert pressure on trade unions within the industry, or it could simply be that the methods used to measure construction productivity are still not adequate for the task. Intuitively, it is difficult to understand how an industry that has seen so many advances in technology, materials and techniques in recent years could actually be experiencing falling productivity. The heterogeneity of construction output remains a complicating factor in productivity measurement, and no doubt there are other factors that need to be considered. Increases in extent and complexity of engineering services is just one of those; greater emphasis on the health and safety of construction workers is surely another. Developments in off-site fabrication also add a layer of difficulty to the problem, as they make measurement of inputs all the more difficult.

The review presented in this chapter is wide ranging: the authors refer to a large number of studies undertaken over many years and describe some of them in detail. As a historical review of construction productivity measurement, it provides a very useful backdrop to some of the newer methodologies that are presented in other chapters in this book.

10 A review of the theory and measurement techniques of productivity in the construction industry

Gerard de Valence and Malcolm Abbott

Introduction

The rate of growth of productivity in Australia, the United Kingdom, the United States and other major economies in the OECD became an issue in the late 1960s, when declining output per hour worked and output per person employed became the focus of a large research programme that sought to interpret and analyse the causes of what became known as the productivity slowdown. At this time, the construction industry's low productivity growth also attracted attention. The rate of growth of productivity of the construction industry has been poor since the 1960s, even by comparison with a long-run overall industry average in the order of 2 to 3 per cent a year.

Despite the efforts made by governments, industry organizations and firms over the past few decades, the rate of measured growth of construction productivity has remained low compared to many other industries. The possible reasons behind this stagnant growth of productivity are various and could include such things as the high labour intensity of the industry, the low economies of scale in the industry, a lack of competition, regulatory impediments, faulty innovation and management practice, poor investment quality and a low levels of skills (Davis 2007). Alternatively, it is possible that the measures used to determine the levels of productivity in the industry, themselves, might be faulty, as was argued in the American case when low levels of productivity growth were detected (see, for instance, Rosefielde and Mills 1979; Schriver and Bowlby 1985).

The purpose of this chapter is to examine some of the issues surrounding the possible stagnation of growth in productivity in the construction industry as well as some of the techniques used in estimating this productivity growth (or possible lack of it). To begin with, a description of the manner in which productivity growth in the construction industry has been estimated in the past is given. In addition, a review is provided of previous research across five areas that have been suggested as important influences on construction productivity.

The different aspects of construction productivity measurement and performance apply at three distinct levels. Which of these three levels is the most appropriate for productivity analysis of construction will depend on the purpose of the analysis. First, at the industry level, the focus is on the measurement of output within the national accounting framework, so we first look at the industry level of measurement of rates of productivity growth. The second level is the heterogeneous nature of construction products, both in type and in location. The chapter collects the limited research on the effects of location and the project-based nature of the industry. Generally, each project is designed and built to serve a special need. Although specific design and construction skills are needed over and over again, the outputs differ in size, configuration, location and complexity. Such uniqueness impacts substantially on construction productivity and the construction process. Third, the site-based nature of construction and project management is discussed. As a subset of these factors, the work-sampling studies carried out on specific tasks, processes or teams should be included. Finally, there is also another set of factors that are here called institutional, and these include construction industry policy, R&D and innovation, technological progress, regulation, the legal framework and procurement and delivery systems.

Productivity measures

The manner in which productivity of the construction industry has been analysed over the years has been influenced by the manner in which productivity analysis more generally has developed. Structural reform in a variety of industries has encouraged researchers to study the productivity and efficiency performance of these industries. In undertaking these examinations, researchers have used a range of productivity and efficiency measurement techniques. In determining the productivity performance of a firm or industry, a range of indicators can be looked at. Utilities and government service providers, such as schools and hospitals, often operate in markets which lack prices and costs determined under competitive conditions. In these cases, the usual market indicators of performance, like profitability and rates of return, cannot be used to gauge a firm or industry's economic performance accurately. It is possible that these financial indicators might be more an indication of the distortions themselves rather than of the performance of the firm or industry in question. In these circumstances, indicators of the level and change of productivity would be more appropriate indicators of performance. Although the construction industry is a relatively competitive one, a number of questions have been raised in the past regarding the degree to which prices and costs in the industry reflect their true economic worth. In these circumstances, therefore, productivity analysis can make a useful contribution to gauging the performance of the industry.

In the past, productivity changes over time were first measured using an 'index' approach. This approach involves the construction of index numbers which can be used to indicate the partial or total factor productivity of an industry. Partial productivity measures generally relate a firm's (or industry's) output to a single input factor – for example, the volume of construction activity per employee is a labour-based partial productivity measure (see, for example, Cassimatis 1969; Briscoe 2006; Cremeans 1981; Ive and Gruneberg 2000: chapter 3; Pearce 2003: chapter 5). Total factor productivity measures are generally the ratio of a total aggregate output quantity index to a total aggregate input quantity index (and total factor productivity growth is then the difference between the growth of the output and input quantity indices).

Partial productivity indicators have the advantage of being easy to compute, requiring only limited data and being intuitively easy to understand. They can, however, be misleading when looking at the change in productivity of a firm or industry. For instance, it might be possible for a company to raise productivity with respect to one input at the expense of reducing the productivity of other inputs. Indices of output to labour, for instance, often tend to overstate the growth of total factor productivity (that is the combined productivity of labour, capital and other factors). Further, capital productivity measures are difficult to calculate given the difficulty in measuring capital inputs and the often very long life of some assets.

Although there are a few studies of industry-level total factor productivity in the construction industry, the literature is not as extensive as it is in many other industries. Most research on construction industry productivity tends to be concerned more with site-level labour productivity, which has a more direct relevance to industry management (see, for instance, Ganesan 1984; Lowe 1987; Maloney 1983; Allen 1985; Thomas *et al.* 1990; Thomas and Sakarcan 1994).

Total factor productivity indices, on the other hand, were first developed at the National Bureau of Economic Research in the United States in the late 1940s. The total factor productivity index approach shows the ratio of an index of combined outputs to an index of combined inputs. The index approaches used to combine outputs and inputs that might be used include the Laspeyres, Paasche and Fisher approaches. A Tornqvist index approach has been used in many total factor productivity studies in recent times.

First of all, labour productivity measurements were undertaken for a range of industries (including construction). This was followed by a range of partial productivity measurements for other inputs such as capital and, in the case of agriculture, land and livestock indicators. A range of studies was then undertaken that attempted to combine the various types of inputs together in a similar fashion to the way in which earlier researchers like Simon Kuznets and Colin Clark had for output in determining levels of gross national product. In undertaking this work, the research by Stigler (1947; 1961), Barton and Cooper (1948), Kendrick and Jones (1951), Schmookler (1952) and Fabricant (1954) was important. Starting with agriculture, this work spread to other individual sectors of the economy as well as nationwide studies. Schmookler (1952) used national-level data to generate nationwide productivity measurements. By the mid-1950s, the National Bureau of Economic Research had published a great deal of work using total factor productivity indices for a range of industries, with one of its employees, John W. Kendrick, publishing extensively using these techniques for a number of years (Kendrick 1956a; 1956b; 1961; 1973). Amongst the industries examined was the construction industry.

Total factor productivity indices of the construction industry in the United States, therefore, were the first part of larger attempts to calculate productivity across the whole economy. Other researchers also began to undertake similar studies of this sort in the 1950s; these included those by Schultze (1959), Kendrick (1961), Haber and Levinson (1956) and Alterman and Jacobs (1961). Over the years, studies have also been conducted in a range of other countries besides the United States using the index approach. They include studies such as those on Hong Kong by Chau (1988), Chau and Lai (1994) and Chau and Walker (1990), for Singapore by Tan (2000), the United States by Stokes (1981) and Allmon *et al.* (2000), the United Kingdom by Briscoe (2006) and New Zealand by Diewert and Lawrence (1999), Black, Guy and McLellan (2003) and Statistics New Zealand (2011). In the studies by Tan (2000) and Black, Guy and McLellan (2003), a slowdown in productivity growth was detected.

The second approach to determining productivity change is econometric measures, which use the estimation of cost or production functions. The estimated functions can then be used to identify changes in productivity or productive efficiency. The estimation of cost functions has been the most commonly used method of determining the levels of efficiency in the industry, although a number of techniques have been used in estimating these cost functions. This approach is also often referred to as the Growth Accounting Approach. Dacy (1965) was perhaps the first to use an estimated production function for the United States construction industry (1947 to 1963) and found increasing levels of productivity. Later examples of estimated production functions include those by Allen (1985), Stokes (1981), Goodrum and Haas (2004) and Kau and Sirmans (1983). Examples of the use of cost functions estimations include those by Schriver and Bowlby (1985) for the United States and Chau (1993 and 2009), who found in both cases rising productivity levels in Hong Kong. Additional work has been undertaken in countries like New Zealand using this approach (Orr 1989; Chapple 1994; Philpott 1995; Mason and Osborne 2007).

Another approach is data envelopment analysis (DEA). The technique was pioneered by Charnes *et al.* (1978) based on the work by Farrell (1957), and there are now many texts offering detailed discussion on DEA, including the algorithms used (see, for example, Knox Lovell and Schmidt 1988; Coelli, Rao and Battese 1998). DEA has been used extensively in a number of industries to assess productivity and efficiency levels (especially utilities). In the case of the construction industry, only a smaller number have been undertaken, and these are looked at in the following chapter.

The accuracy and impact of price indices

Most studies that look at the productivity of the construction industry are centred on the industry in the United States. One element some researchers found was that of stagnation or declining productivity in the United States construction industry, and these researchers had some difficulty explaining why this might have occurred. These include work by Stokes (1981), Allen (1985) and Schriver and Bowlby (1985). Tan (2000) found a similar decline in productivity in the Singapore construction industry over the period 1980 to 1996.

One of the reasons the measured rate of construction productivity growth may be low is because of the measurement of output as value added (the total value of goods and services produced after deducting the costs in the production process) is adjusted by a deflator for movements in prices. The construction deflator may not fully take these movements into account, and therefore real output is underestimated. Also, the significant role of changes in the quality of construction may not have been rigorously measured and reflected in changes in real value added.

Output of the construction industry is estimated by deflating current price figures by input price indices. A number of researchers have criticized the use of input price indices for deflating construction expenditure, for being unrepresentative of the inputs priced and geographical coverage and for being based on inaccurate weights. The Stigler Report (1961:29) recommended a significant increase in research on construction deflation and suggested a residential deflator based on the price per square foot of a range of categories of new homes. This led, in 1968, to the adoption of a new, hedonic price index for housing in the United States by the Bureau of Economic Analysis.

A number of alternative deflators have been developed. Allen (1985) used a price-per-square-foot index for deflating nonresidential building, assuming that this is a good proxy for output. According to Allen's estimates, about half the decline in construction productivity during the 1960s and 1970s was due to the overdeflation of construction output. Cassimatis found that price indices cannot provide adequate deflators for construction: 'the feeling persists that construction productivity is greater than the measurements show. . . largely due to the fact that there are no adequate price indices that can be used as deflators of the gross product' (Cassimatis 1969:79–80).

Pieper (1990) also argued that deflation by input price indices does not produce suitable estimates of output at constant prices and, given the extensive use of input price indices as deflators in estimating the constant price of output for the construction industry, productivity measurement for this industry is problematic, to say the least. Pieper concluded that, for the United States: "evidence indicates an over deflation of construction of at least 0.5% per year between 1963 and 1982".

Chau and Lai (1994) developed a system for measuring the relative labour productivity of the Hong Kong construction industry. Their approach used a method of measuring the relative labour productivity of the industry from national accounts data, and then derived the trend of construction labour productivity. This discussion of relative rates of growth of labour productivity used an implicit price deflator for net output of the construction industry obtained through double deflation but did not discuss the nature of the price indices used or their applicability. The price indices were based on a construction output price index and a material cost index using the methodology developed by Chau and Walker (1988).

Lowe (1995) described the use of estimation indices by Statistics Canada, using surveys sent to subcontractors. Around 100 different items were priced for five building types, and each of five types had its own index. A recent analysis of British building price indices by Yu and Ive (2008) found that these indices measure the price movement of the traditional building trades but almost completely ignore mechanical and electrical services.

Cannon (1994) questioned the accuracy of contractor statistics and Briscoe (2006) asked, "How useful and reliable are construction statistics?" These papers identified a range of problems with data collection and analysis, including defining the scope and coverage of the industry, measuring outputs across different types of activity, identifying construction firms, measuring capital formation and capital stock and inconsistent employment statistics. Crawford and Vogel (2006) also drew attention to data limitations for productivity analysis.

Regional and sectoral effects on productivity

As well as problems associated with choosing the best possible price indices, other hypotheses also exist that attempt to explain the decline in construction productivity. Some common ones include that there has been a decline in the capital–labour ratio in the industry (Blake *et al.* 2004), changes in the age–sex composition of the labour force (Cremeans 1981), a shift towards nonunion construction (Allen 1985), an increase in government regulation (Tucker 1986) or cyclical and business cycle effects. Project characteristics such as the increased size and complexity of projects, resulting communication difficulties and fast-tracking projects where design and construction phases overlap also affect coordination. There have been a few papers that address the effects of these on productivity (for examples, see Table 10.1).

Cremeans (1981) discussed a number of hypotheses that had been proposed to explain the significant decline in construction industry labour productivity in the 1970s. Only one of the hypotheses, the increased proportion of younger, less experienced workers, was supported by the available data. Bowlby and Schriver's (1986) analysis of United States productivity data indicated seven compositional changes in building, and they suggested that these would account for much of the productivity slowdown.

Table 10.1 Representative papers: regional and sectoral effects on industry productivity

Cremeans (1981)	Found younger, less experienced workers the main cause
Bowlby and Schriver (1986)	Identified seven compositional changes in building, and these account for much of the productivity slowdown
Tucker (1986)	The increased size and complexity of construction projects
Ive *et al.* (2004)	The output structure of a country's construction industry will influence average labour productivity
Blake *et al.* (2004)	UK construction has lower capital per worker than France, Germany and the US

Project-based nature of the industry and the role of project management

A large number of papers have recommended that construction productivity could be improved through the use of flexible organization structures, favourable union attitudes, higher worker motivation and improved overtime and change order strategies (examples of these studies can be found in Table 10.2). Most of these surveys found that cost control, scheduling, design practices, labour training and quality control are the functions that are consistently seen as having room for improvement. Often the fragmented nature of the industry is seen as a hindrance to improving productivity (Ganesan 1984). However, Chau and Lai (1994) suggest that productive efficiency is increased by the division of labour.

Borcherding (1976) identified the factors having an adverse effect on construction productivity as union attitudes, worker selection practices and motivation, inflexible bureaucratic organization structures, overtime and change orders. Using these factors, Herbsman and Ellis (1990) developed a statistical model of the quantitative relationships between influence factors and productivity rates.

Koehn and Brown (1986) argued that construction productivity is affected by a wider range of variables, which they divided into the six areas of management, labour, government, contracts, owner characteristics and financing. Koehn and Caplan (1987) focused on small to medium-sized construction firms rather than large construction firms. Their study concluded that productivity improvement efforts should be concentrated in planning, scheduling, site and labour management functions. Laufer and Jenkins (1982) also focused on the management issues and discussed them in the context of motivation of workers. They suggested that while motivation does not directly influence the rate of working, motivation directly impacts the percentage of working time spent productively.

Table 10.2 Representative papers: project-based nature of the industry and the role of project management

Borcherding and Oglesby (1974)	Concluded well-organized construction jobs which permit workers to be productive lead directly to job satisfaction
Borcherding (1976)	Identified six factors having adverse effects on construction productivity
Kellogg *et al.* (1981)	Argued that the fragmented nature of the industry impedes productivity growth
Ganesan (1984)	Also argued fragmentation affects productivity
Hague (1985)	Found financial incentives and any other method for encouraging productivity have had arguments for and against
Koehn and Caplan (1987)	Productivity improvement efforts should be concentrated on planning, scheduling, supervision, and labour
Briscoe (1988)	The quality of construction management is an important factor which helps to explain low productivity
McFillen and Maloney (1988)	Found contractors did little to encourage good performance, so workers reported little incentive to be highly productive
Herbsman and Ellis (1990)	Developed a statistical model of quantitative relationships between influence factors and productivity rates
Chau and Lai (1994)	Argue the fragmented nature of the industry is often seen as a hindrance to improving productivity
Dai, Goodrum and Maloney (2007)	Foremen reported project management factors having more impact on their productivity, and craft workers reported factors related to construction materials having more impact

Arditi and Mochtar's surveys of the top 400 United States contractors in 1979, 1983 and 1993 identified areas with potential for productivity improvement. The functions needing more improvement in 1993 compared with the previous survey were prefabrication, new materials, value engineering, specifications, labour availability, labour training and quality control, whereas those that were identified as needing less improvement were field inspection and labour contract agreements (Arditi and Mochtar 2000).

Allmon *et al.* (2000) presented an approach to long-term productivity trends in the United States construction industry over the past 25 to 30 years. Means' cost manuals (the main United States source of estimating data) were used to trace the values for tasks undertaken in the process of construction, and changes in these values were taken as productivity trends. Unit labour costs in constant dollars and daily output factors were compared over decades for each task. Direct work rate data from 72 projects in Austin, Texas over the last 25 years was also examined. The combined data indicated that productivity had increased in the 1980s and 1990s. Depressed real wages and technological advances appear to be the two biggest reasons for this increase. Their data also indicated that management practices were not a leading contributor to construction productivity changes over time.

Procurement systems and the effectiveness of construction industry policy and intervention

Some researchers have identified institutional factors responsible for construction productivity levels. Labour issues include organized labour, the competencies of project participants, the tendency of site management to spend more time providing information and writing reports than actually managing the project and the inadequacies of an educational system which produces graduates with excellent skills in analysis and design but with little knowledge of methods to turn designs into realities (Tucker 1986). Other institutional issues are the tendency of construction firms to become larger and more specialized, legal restrictions on the management of construction projects and the complex regulatory regimes the industry works under (Table 10.3).

Table 10.3 Representative papers: procurement and delivery systems and the effectiveness of construction industry policy

Cassimatis (1969)	The major factors affecting the efficiency of organizations in construction are institutional
Tucker (1986)	Institutional issues were the tendency of construction firms to become larger and more specialized, legal restrictions on the management of construction projects and insufficient research in construction and project management methods
Sidwell (1987)	Described the Australian construction industry as thoroughly conservative and slow to change in any fundamental way
Cox and Townsend (1998)	Construction has not developed the supply chains and procurement methods other industries have
Craig (2000)	Compared traditional and D&B (design and build) procurement for innovation
Dubois and Gadde (2002)	The separation of design and construction creates inefficiencies

The limitations of the traditional procurement method have contributed to the poor performance of the construction industry and have prompted the development of alternative procurement strategies designed to facilitate improvements in the way buildings and structures are delivered (Cox and Townsend 1998). Craig (2000) concluded that the traditional tendering process for building works does not encourage design innovation by tenderers, because tendering rules produce direct price competition for a specified product.

R&D, innovation and productivity

The construction industry has not established an impressive track record in innovation or technical advancement. The main effort in industry development has been concentrated in procurement, planning and management and design improvements. Nevertheless, there have been some significant advances in construction technology over the last two decades in both the materials used and the application of new construction methods (Fairclough 2002).

Gann (2003:554) cites Bowley (1960) as showing that construction is an adopter of innovations from other industries rather than a source of innovation. Bowley's work 'shows that demand for new types of buildings is usually more important in stimulating radical technical and organizational innovation than the need to erect better and cheaper buildings to accommodate existing functions'. Cassimatis (1969) concluded his study with a chapter on institutional factors, because 'once the contract is awarded, competitive forces do not always prevail' (Cassimatis 1969:118). Institutional factors that affect the performance of the industry are its openness to innovation and capturing of economies of scale.

Koch and Moavenzadeh (1979) focused on the role of technology in highway construction and found there had been substantial gains in both labour and capital productivity over the previous 50 years in the United States. They concluded that future gains in efficiency can be expected to be less than the previous gains, so new means of accomplishing technological change in the construction industry are needed. Arditi (1985) conducted a study of large construction firms to determine potential areas for construction productivity improvement. One of the study's conclusions was that more productive construction technology such as industrialized building processes is important in achieving higher levels of construction productivity.

Hobday (1998) and Gann and Salter (2000) argued that the construction industry can and should be more innovative. Many papers follow Tatum's (1986) analysis of the industry in terms of advantages and constraints to innovation, and despite the Tatum model of construction innovation being more than two decades old, it still captures many of the key features of the discussion raised by more recent efforts such as Reichstein *et al.* (2005), Fairclough (2002) or Slaughter (1998). Ivory (2005) suggested that clients of builders will not be prepared to pay for innovation.

Table 10.4 Representative papers: contribution of research and development and innovation

Rosefielde and Mills (1979)	Argue the rate of technological progress in the construction industry may be slow because buildings are heterogeneous
Koch and Moavenzadeh (1979)	Focused on the role of technology in highway construction and concluded new means of accomplishing technological change are needed
Arditi (1985)	Recommended areas that research should concentrate on
Tatum (1986)	Construction has many features that favour innovation
Gann (1997)	Discusses the role of government-funded R&D
Gann and Salter (2000)	Construction has the potential to be more innovative
Fairclough (2002)	Construction lags in R&D and innovation
Hobday (1998)	Argues that the nature of construction projects and teams creates opportunities for innovation
Zhi, Hua, Wang and Ofori (2003)	Seven factors influencing TFP growth in the construction industry of Singapore over 1984–1997 were identified
Ivory (2005)	Argued clients will avoid risk associated with innovation

Conclusion

The rate of growth of productivity in the construction industry in a number of countries has lagged behind that of other industries for at least five decades, and the earliest studies that identified this problem date from the late 1960s in the United States with Cassimatis's (1969) analysis of labour productivity growth in construction between 1947 and 1967. This is despite there having been a range of technological changes that have occurred in the industry, such as the introduction of new handheld powered tools, improved lifting and moving machinery and new materials and processes.

Two possible explanations for the lack of demonstrable improvement in construction productivity are possible. The first is the importance of measurement and data to the research. This belongs to a broader set of issues about the structure and use of price indices in the national accounting framework, an area where construction economists might have an opportunity to make a contribution. Recently there has been a shift from the use of deflators and their effects on measured output (or, more precisely, the ratio of output to labour input) to concern over the boundaries of the production system and more accurate measurement of specific factors such as capital inputs adjusted for quality and employment adjusted for firm size.

The second is the diversity of other issues raised that are suggested as affecting productivity. Influences on productivity growth in the construction industry, apart from the nature of the product, can be traced to the nature of the methods used in delivering and managing the processes involved. Construction is a labour-intensive industry in comparison with manufacturing industries, but there has been a significant increase in the prefabricated component of construction, which could have been expected to lead to productivity growth. Also, construction methods have tended to become more capital intensive as the number of cranes and the variety of equipment and hand tools used has increased. However, the productivity growth that one would expect to observe as a result of these trends has not occurred, according to measurements of productivity growth by the major national statistical agencies and reports like the UK studies by Ive *et al.* (2004) and Blake *et al.* (2004).

This chapter has reviewed a wide range of previous research addressing a range of factors that could affect productivity. The bringing together of these different literatures on productivity analysis and measurement, project procurement and delivery systems, construction industry policy and intervention, and R&D and innovation allows a broader perspective on the construction industry's productivity performance. In terms of applicability, the breadth of management issues raised by researchers points to some possibly serious problems with both the management of projects and the management of workers. After several decades of development of project management techniques, the average performance of projects does not appear to have improved greatly, with the more recent research finding problems similar to those found in the early work. Last, it is possible that the R&D profile of the industry is as much an artefact of the data as a real problem. Construction is an industry that readily adopts research developments in other industries, the use of computers and the constant flow of new products from manufacturers supplying materials and equipment being good examples. R&D expenditure within the industry will not be very high in this case.

References and Further Reading

Allen, S. G. (1985) Why construction industry productivity is declining. *Review of Economic Statistics*, **117** (4), 661–665.

Allmon, E., Haas, C., Borcherding, J. and Goodrum, P. (2000) US Construction Labor Productivity Trends 1970–1998. *Journal of Construction Engineering and Management*, ASCE, **126** (2), 97–104.

Alterman, J. and Jacobs, E. E. (1961) Estimates of real product in the United States by industrial sector, 1947–1955. *Output, Input and Productivity Measurement*.

National Bureau of Economic Research (Cambridge, MA: Princeton University Press).

Arditi, D. (1985) Construction productivity improvement. *Journal of Construction Engineering and Management*, ASCE, **111**, 1–14.

Arditi, D. and Mochtar, K. (2000) Trends in productivity improvement in the US construction industry. *Construction Management and Economics*, **18** (1), 15–27.

Barton, G. T. and Cooper, M. R. (1948) Relation of agricultural production to inputs. *Review of Economics and Statistics*, **30** (7), 117–126.

Black, M., Guy, M. and McLellan, N. (2003) *Productivity in New Zealand 1988 to 2002*. Treasury Working Paper 03/06 (Wellington: The Treasury).

Blake, N., Croot, J. and Hastings, J. (2004) *Measuring the Competitiveness of the UK Construction Industry, Vol. 2: Industry Economics and Statistics* (London: Department of Trade and Industry).

Borcherding, J.D. (1976) Improving productivity in industrial construction. *Journal of Construction Division* ASCE, **102** (C04), 599–614.

Borcherding, J.D. and Oglesby, C.H. (1974) Construction productivity and job satisfaction. *Journal of Construction Division* ASCE, **100** (3), 413–431.

Bowlby R. and Schriver, W. (1986) Observations on productivity and composition of building construction output in the United States, 1972–1982, *Construction Management and Economics*, **4** (1), 1–18.

Bowley, M. (1960) *Innovation in Building Materials* (London: Gerald Duckworth).

Briscoe, G. (2006) How useful and reliable are construction statistics? *Building Research & Information*, **34** (3), 220–229.

Cannon, J. (1994) Lies and construction statistics. *Construction Management and Economics*, **12** (4), 307–312.

Cassimatis, P.J. (1969). *Economics of the Construction Industry* (New York: National Industrial Conference Board).

Chapple, S. (1994) *Searching for the Heffalump? An Exploration into Sectoral Productivity and Growth in New Zealand*. Working Paper 99/10 (Wellington: New Zealand Institute of Economic Research).

Charnes, A., Cooper, W.W. and Rhodes, E. (1978) Measuring the efficiency of decision making units. *European Journal of Operational Research*, **2** (6), 429–444.

Chau, K.W. (1990) *Total Factor Productivity of the Building Industry of Hong Kong*. University of Hong Kong, Hong Kong: Unpublished PhD thesis.

Chau, K.W. (1993) Estimating industry-level productivity trends in the building industry from building cost and price data. *Construction Management and Economics*, **11** (4), 370–383.

Chau, K.W. (2009) Explaining total factor productivity trend in building construction: empirical evidence from Hong Kong. *International Journal of Construction Management*, **9** (2), 45–54.

Chau, K.W. and Lai, L.W.C. (1994) A comparison between growth in labour productivity in the construction industry and the economy. *Construction Management and Economics*, **12**, 183–185.

Chau, K.W. and Walker, A. (1988) The measurement of total factor productivity of the Hong Kong construction industry. *Construction Management and Economics*, **6** (3), 209–224.

Chau, K. W. and Walker, A. (1990) Industry-level Productivity Trend of the Construction Industry—Some Empirical Observations in Hong Kong. *Proceedings of the 1990 CIB International Symposium on Building Economics and Construction Management*, Vol. 1, 79–90.

Chau, K. W. and Wang, Y. S. (2005) An analysis of productivity growth in the construction industry: a non-parametric approach. In: Khosrowshahi, F. (ed.), *21st Annual ARCOM conference*, 7–9 September SOAS, University of London (Association of Researchers in Construction Management), 1, 159–169.

Coelli, T., Prasada Rao, D. S. and Battese, G. E. (1998) *An Introduction to Efficiency and Productivity Analysis* (Boston: Kluwer).

Cox, A. and Townsend, M. (1998) *Strategic Procurement in Construction* (London: Thomas Telford Publishing).

Craig, R. (2000) Competitive advantage through tendering innovation. *Australian Construction Law Newsletter*, 70, March (Sydney: UTS Design, Architecture and Building).

Crawford, P. and Vogel B. (2006) Measuring productivity in the construction industry. *Building Research & Information*, 34 (3), 208–219.

Cremeans, J. E. (1981) Productivity in the construction industry. *Construction Review*, 27 (5), 4–6.

Dacy, D. C. (1965) Productivity and price trends in construction since 1947. *Review of Economics and Statistics*, 47 (4), 406–411.

Dai, J., Goodrum, P. M. and Maloney, W. F. (2007) Analysis of craft workers' and foremen's perceptions of the factors affecting construction labour productivity. *Construction Management and Economics*, 25 (11), 1139–1152.

Davis, N. (2007) *Construction Sector Productivity Scoping Report* (Wellington: Martin Jenkins).

Diewert, E. and Lawrence, D. (1999) *Measuring New Zealand's Productivity*. Treasury Working Paper 99/5 (Wellington: The Treasury).

Dubois, A. and Gadde, L-E. (2002) The construction industry as a loosely coupled system: implications for productivity and innovation. *Construction Management and Economics*, 20 (7), 621–632.

Fabricant, S. (1954) *Economic Progress and Economic Change* (Washington, DC: National Bureau of Economic Research).

Fairclough, J. (2002) *Innovation in the Construction Industry – a Review of Government R&D Policies and Practices* (London: Department of Trade and Industry).

Färe, R., Grosskopf, S. and Margaritis, D. (1996) Productivity growth. In: Silverstone, B., Bollard, A. and Lattimore, R. (eds.), *A Study of Economic Reform: the Case of New Zealand* (New York: Elsevier Science).

Farrell, M. (1957) The measurement of productive efficiency. *Journal of the Royal Statistical Society*, Series A, **CXX**, 253–281.

Ganesan, S. (1984) Construction productivity. *Habitat International*, 8, 3–4.

Gann, D. M. (1997) Should governments fund construction research? *Building Research and Information*, 25, 257–267.

Gann, D. M. (2003) Guest editorial: innovation in the built environment. *Construction Management and Economics*, 21, 553–555.

Gann, D. M. and Salter, A. J. (2000) Innovation in project-based, service-enhanced firms: the construction of complex products and systems. *Research Policy*, 29, 955–972.

Goodrum, P. M. and Haas, C. T. (2004) Long-term impact of equipment technology on labor productivity in the U.S. construction industry at the activity level. *Journal of Construction Engineering and Management*, 130 (1), 124–133.

Haber, W. and Levinson, H. (1956) *Labor Relations and Productivity in the Building Trades* (Ann Arbor: University of Michigan Press).

Hague, D. J. (1985) Incentives and motivation in the construction industry: a critique. *Construction Management and Economics*, 3 (2), 163–170.

Herbsman, Z. and Ellis, R. (1990) Research of factors influencing construction productivity. *Construction Management and Economics*, 8 (1), 49–61.

Hobday, M. (1998) Product complexity, innovation and industrial organisation. *Research Policy*, 26, 689–710.

Ive, G. and Gruneberg, S. (2000) *The Economics of the Modern Construction Sector* (Basingstoke: Palgrave Macmillan).

Ive, G., Gruneberg, S., Meikle, J. and Crosthwaite, D. (2004) *Measuring the Competitiveness of the UK Construction Industry* (London: Dept. of Trade and Industry).

Ivory, C. (2005) The cult of customer responsiveness: is design innovation the price of a client-focused construction industry? *Construction Management and Economics*, 23, 861–870.

Kau, J. B. and Sirmans, C. F. (1983) Technological change and economic growth in housing. *Journal of Urban Economics*, 13, 283–295.

Kellogg, J. C., Howell, C. G. and Taylor, D. C. (1981) Hierarchy model of construction productivity. *Journal of the Construction Division* ASCE, 161 (38), 137–152.

Kendrick, J. W. (1956a) *Productivity Trends: Capital and Labor*. Occasional Paper 53 (Washington DC: National Bureau of Economic Research).

Kendrick, J. W. (1956b) *The Meaning and Measurement of National Productivity*. George Washington University, Washington DC: PhD thesis.

Kendrick, J. W. (1961) *Productivity Trends in the United States* (New York: National Bureau of Economic Research/Princeton University Press).

Kendrick, J. W. (1973) *Postwar Productivity Trends in the United States 1948–1966* (New York: National Bureau of Economic Research).

Kendrick, K. W. and Jones, C. E. (1951) Gross national farm product in constant dollars, 1910–1950. *Survey of Current Business*, 31, September, 12–19.

Koch, J. A. and Moavenzadeh, F. (1979) Productivity and technology in construction. *Journal of the Construction Division*. Proceedings of the American Society of Civil Engineers, 105 (C04), December, 351–366.

Koehn, E. and Brown, G. (1986) International Labor Productivity Factors. *Journal of Construction Engineering and Management*, 112 (2). DOI: 10.1061/(ASCE)0733-9364(1986)112:2(299)

Koehn E. and Caplan, S. B. (1987) Work improvement data for small and medium size contractors. *Journal of Construction Engineering and Management*, 113 (2), 327–339.

Knox Lovell, C. A. and Schmidt, P. (1988) A comparison of alternative approaches to the measurement of productive efficiency. In: Dogramaci A. and Färe, R. (eds.), *Applications of Modern Production Theory: Efficiency and Productivity* (Boston: Kluwer).

Laufer, A. and Jenkins, G. (1982) Motivating construction workers, *Journal of the Construction Division*, 108 (CO4) 531–545.

Lowe, J. G. (1987) The measurement of productivity in the construction industry. *Construction Management and Economics*, 5, 115–121.

Lowe, P. (1995) Labour-productivity growth and relative wages: 1978–1994. In: Andersen, P., Dwyer, J. and Gruen, D. (eds.), *Productivity and Growth: Proceedings of a Conference* (Sydney: Reserve Bank of Australia).

Maloney, W. (1983) Productivity improvement: the influence of labor. *Journal of Construction Engineering Management*, 109 (3), 321–334.

Mason, G. and Osborne, M. (2007) *Productivity, Capital-Intensity and Labour Quality at Sector Level in New Zealand and the UK.* New Zealand Treasury Working Paper 07/01 (Wellington: The Treasury).

McFillen, J.M. and Maloney, W.F. (1988) New answers and new questions in construction worker motivation. *Construction Management and Economics,* 6 (1), 35–48.

Orr, A. (1989) *Productivity Trends in New Zealand: a Sectoral and Cyclical Analysis 1961–1987* (Wellington: NZIER).

Pearce, D. (2003) *The Social and Economic Value of Construction* (London: Davis Langdon Consultancy).

Philpott, B. (1995) *New Zealand's Aggregate and Sectoral Productivity Growth 1960–1995.* Research Project on Economic Planning, Paper 274 (Wellington: Victoria University of Wellington).

Pieper, P.E. (1990) The measurement of construction prices: retrospect and prospect. In: Berndt E.R. and Triplett J.E. (eds.), *Fifty Years of Economic Measurement* (Chicago: National Bureau of Economic Research, University of Chicago Press).

Reichstein, T., Salter, A.J. and Gann, D.M. (2005) Last among equals: a comparison of innovation in construction, services and manufacturing in the UK. *Construction Management and Economics,* 23, 631–644.

Rosefielde, S. and Mills, D.Q. (1979) Is construction technologically stagnant? In: Lange, J.E. and Mills, D.Q. (eds.), *The Construction Industry* (Lexington, MA: D.C. Heath and Company).

Schmookler, J. (1952) The changing efficiency of the American economy, 1869–1938. *Review of Economics and Statistics,* 34 (3), 214–231.

Schriver, W.R. and Bowlby, R.L. (1985) Changes in productivity and composition of output in building construction, 1972–1982. *Review of Economics and Statistics,* 67 (2), 318–322.

Schultze, C.L. (1959) *Prices, Costs and Output for the Post War Decade: 1947–1957* (New York: New Committee for Economic Development).

Sidwell, A.C. (1987) *The Future for Building Education.* Occasional Paper No. 3 (Canberra: The Australian Institute of Building).

Slaughter, S. (1998) Models of construction innovation. *Journal of Construction Engineering and Management,* 124 (30), 226–231.

Statistics New Zealand (2011) *Industry Productivity Statistics, 1978–2009* (Wellington: Statistics New Zealand).

Stigler, G.J. (1947) *Trends in Output and Employment* (Cambridge, MA: National Bureau of Economic Research, University of Princeton Press).

Stigler, G.J. (1961) Economic problems in measuring changes in productivity. In: *Output, Input and Productivity Measurement,* Conference on Research in Income and Wealth, Cambridge, MA: National Bureau of Economic Research, University of Princeton Press.

Stigler Report (1961) *Price Statistics of the Federal Government: Review, Appraisal and Recommendations* (New York: Price Statistics Review Committee, NBER).

Stokes, H.K. (1981) An examination of the productivity decline in the construction industry. *Review of Economics and Statistics,* 63 (4), 495–502.

Tan, W. (2000) Total factor productivity in Singapore construction. *Engineering, Construction and Architectural Management,* 7 (2), 154–158.

Tatum, C.B. (1986) Potential mechanisms for construction innovation. *Journal of Construction Engineering and Management* ASCE, 112 (2).

Thomas, H., Maloney, W., Horner, R., Smith, G., Handa, V. and Sanders, S. (1990) Modelling construction labor productivity. *Journal of Construction Engineering Management*, **116** (4), 705–726.

Thomas, H. and Sakarcan, A. (1994) Forecasting labor productivity using factor model. *Journal of Construction Engineering Management*, **120** (1), 228–239.

Tucker, R. L. (1986) Management of construction productivity. *Journal of Management Engineering*, **2** (3), 148–156.

Wang, Y. S. (1998) *An Analysis of the Technical Efficiency in Hong Kong's Construction Industry*. University of Hong Kong, Hong Kong: Unpublished PhD thesis.

Wang, Y. S. and Chau, K. W. (1997) An evaluation of the technical efficiency of construction industry in Hong Kong using the DEA approach. *Proceedings of ARCOM97 Conference*. Cambridge, UK, 690–701.

Wang, Y. S. and Chau, K. W. (2001) An assessment of the technical efficiency of construction firms in Hong Kong. *International Journal of Construction Management*, **29**, 105–122.

Yu, M.K.W. and Ive, G. (2008) The compilation methods of building price indices in Britain: a critical review. *Construction Management and Economics*, **26** (7), 693–705.

Zhi, M., Hua, G. B., Wang, S. Q. and Ofori, G. (2003) Total factor productivity growth accounting in the construction industry of Singapore. *Construction Management and Economics*, **21**, 707–718.

Editorial comment

Neurons are nerve cells, specialized cells in living creatures that transmit and process information. Neurons are connected in groups called neural networks that operate in a coordinated manner and allow sophisticated organisms such as human beings to perform all sorts of intricate actions. A human brain has something in the order of 100 billion (10^{11}) of these cells and as many as 100 trillion (10^{14}) connections (*synapses*) that connect them to one another (Williams and Herrup 1988).

It is this abundance of processors that allows us to perform routine yet highly complex tasks such as hitting a moving object with a stick with little or no conscious thought, with neurons processing and reprocessing information on trajectory, velocity and other variables at extraordinary speed. A newborn baby does not have this sort of capacity, but babies learn and develop these skills over time.

Computer scientists have developed artificial neural networks (ANNs) that are modelled on the behaviour of the natural neural networks that exist in the human brain and nervous system. A key characteristic of these ANNs is that, like those in nature, such networks can be trained to recognize patterns and to make decisions based on what they detect. To date, ANNs remain far smaller than the networks found in humans, but they are the focus of ongoing research, particularly amongst those who are trying to develop artificial intelligence.

One specialized area of research in the field of ANNs is their application in the construction industry; in this chapter, some recent research into their application in relation to the assessment of construction productivity and as a tool for estimating construction cost is described. Expert systems that utilize the processing power of computers have been developed over several decades, but the key difference in those based on ANNs is the ability of these systems to 'learn' more effectively from 'experience'. Expert systems that can learn from experience have been around for quite some time; computers that can play chess are a good example. An ANN, however, should have the capacity to generalize rather than to merely remember previous situations that the system has encountered. This concept is at the heart of the search for artificial intelligence, and it is also central to what sets humans apart from other living creatures. It is the key to the so-called Turing Test that determines whether a machine can imitate human behaviour to the extent that a person conversing with a machine cannot distinguish the machine's responses from those of a human being. This obviously goes beyond simply responding to particular questions with a preprogrammed answer.

In this chapter, the authors provide an introduction to the development and use of ANNs in construction.

11 Construction productivity and cost estimation using artificial neural networks

Ali Najafi and Robert Tiong

Introduction

Because of the uncertainties and complexities involved in construction projects, expert system applications and artificial intelligence are helpful in the context of construction engineering and management. This chapter focuses on the applications of artificial neural networks (ANNs) for productivity and cost estimations, which are among the most crucial tasks of construction managers and general estimators.

The main objective of this chapter is to provide practical explanations of how to design, develop, analyse and validate ANNs as robust and reliable tools for productivity and cost estimations. An introduction to ANNs is provided, and several examples from the literature that have used ANNs in different areas of construction productivity and cost predictions are listed. As a result, a framework is presented to serve as a general guide on how to develop ANNs, and based on that, a detailed example is discussed to show the application in a real construction project setting. By the end of the chapter, readers should have some basic background about ANNs and should be able to develop a simple but efficient ANN for their own construction projects.

Artificial neural networks (ANNs)

An artificial neural network (ANN) can be defined as a massive parallel distributed processor composed of simple processing units (neurons) which are capable of storing experiential knowledge and retrieving it for future use (Haykin 1999). Neurons communicate by sending signals to each other over a large number of weighted connections. Thus, ANNs can be considered as an information processing technology that, by learning from different experiences and generalizing from previous examples, can simulate the human brain system. A simple schematic diagram of a neuron is shown in Figure 11.1.

Figure 11.1 shows that each neuron has two distinct segments: a summing junction that sums up the received inputs from neighbours and an activation function that computes the output signal, which is propagated to other

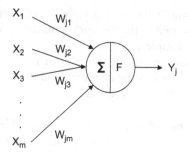

Figure 11.1 Schematic diagram of a neuron

neurons. The activation function can be theoretically in any form such as signum, linear or semilinear, hyperbolic tangent and sigmoid functions.

To form a network, neurons are grouped into several layers, namely input, hidden and output layers. Two types of network topologies are shown in Figure 11.2:

- Feed-forward networks: data flows strictly from input to output layers, and no feedback connections are allowed.
- Recurrent networks: feedback connections are allowed to provide data flow from the following layers to the preceding layers.

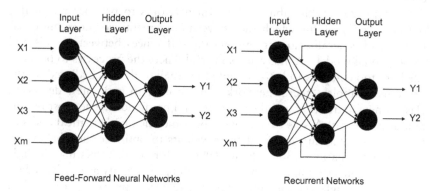

Figure 11.2 Different topologies of ANNs (Alavala 2006)

An ANN should be arranged in such a way that it can provide the desired outputs for a set of inputs presented to the network. To do so, either connection weights should be set using prior knowledge or the network should be trained by training samples (training sets). Connection weights are then updated based on 'the learning rules'. There are two types of learning in ANNs:

- Supervised learning (associative learning): both input and output training pairs are presented to the network, and the network will learn based on the presented samples.
- Unsupervised learning (self-organization): there is no prior set of categories to be presented to the network, and the system develops its own representation of the input stimuli.

Almost all the learning rules are variants of the 'Hebbian' learning rule, which simply says that if two neurons are active, their interconnection must be strengthened. Based on this learning rule, the 'delta rule' uses the differences between actual and desired activation for modifying the connection weights.

Multilayer feed-forward neural networks (Figure 11.2) are the most commonly used architecture in the engineering fields, including construction engineering and management. The so-called back-propagation learning algorithm (Rumelhart, Hintont *et al.* 1986), which is a generalization of the delta rule, is generally used in these types of networks. A set of training samples is presented to the network, and based on the activation functions of the hidden and output layers, the network outputs are calculated, and are then compared with the actual (desired) outputs. The error is calculated (i.e. the difference between the actual and network output) and is propagated back to the network, the connection weights are modified and this procedure is repeated until the network reaches the stop criterion.

Generalization is the ability of neural networks to produce reasonable outputs for inputs for which they have not been trained. Learning capabilities and generalization are among the differences between ANNs and conventional models. ANNs are also useful where the processes to be modelled are too complex to be represented by analytical models. Based on the previous examples, neural networks can be built to make new decisions, classify new patterns and make forecasts and predictions. The most important advantage of ANNs over conventional models (mathematical and statistical) is their ability to adjust their weights to optimize their behaviour in different situations (Boussabaine 1996). Moreover, ANNs may include multiple outputs simultaneously (e.g. cost and productivity are both predictable using a single network providing the required inputs are used) when compared to multiple regression analysis (MRA), which can only estimate one output at a time.

Using ANNs for productivity and cost estimations

This section describes the procedure for the development of simple but efficient ANNs for productivity and cost predictions. Past studies are described and a framework established that shows the steps required for the development of ANNs for productivity and cost estimations. Based on that, an example is provided to show the accuracy of such networks in the context of real construction projects.

Suppose an estimator wants to predict the total construction cost of a new high-rise building project for a general contractor. The focus here is to describe the problem in such a way that it can be solved by the predictive ability of an ANN. To do so, the first step is to collect the factors (inputs) that affect the cost of the project, which are mainly project specifications such as total gross floor area, the type of building structure (steel or concrete), specific resources required and the like. Note that the main output is the total cost of the building. The estimator is then required to look at the previous projects (similar projects by the same contractor or other general contractors) to be used as the training examples for the ANN. The next step is to configure the network by choosing the number of layers, type of activation functions for hidden and output layers and number of neurons in each layer. These configurations are usually determined by trial and error. It has been shown that one layer of hidden units is enough to approximate any function to an arbitrary level of precision (Alavala 2006), but there is no specific rule regarding the number of neurons in the hidden layer.

There are several commercial packages available for design and analysis of ANNs. Most of them have graphical interfaces and can be used as add-ins for commonly used spreadsheet programs like Microsoft Excel. After the network is developed and trained, it is ready to be used for estimation purposes in the new projects with acceptable accuracy considering the uncertainties involved in the construction industry. A similar procedure can be used to estimate the productivity of different construction operations.

The following are some examples from the literature that have used the same method to develop ANNs for productivity and cost estimation of different construction projects. Some of the following examples have used other learning algorithms, but the procedure to develop the ANNs is more or less the same.

ANNs for total cost and duration estimations

One of the most important tasks for the construction management team is to provide cost and schedule estimations for different construction projects. Because of the predictive capabilities of ANNs, they have been widely used to provide more accurate cost and schedule predictions when compared to conventional estimation methods. This can result in fewer cases of cost overruns and fewer schedule discrepancies during the construction phase. Some of the examples in this area included predicting:

- highway and road tunnel construction costs (Hegazy and Ayed 1998; Wilmot and Mei 2005; Wichan *et al.* 2009; Petrousatou *et al.* 2011)

- duration of reconstruction projects and the related costs (Attalla and Hegazy 2003; Chen and Huang 2006)
- total cost and construction duration of residential buildings (Bhokha and Ogunlana 1999; Emsley *et al.* 2002; Kim *et al.* 2004)
- cost of structural systems of buildings (Günaydın and Doğan 2004)
- cost indices and unit price analysis (Baalousha and Çelik 2011).

The list shows that ANNs can be applied at any level for cost and duration estimations, from unit price analysis, which is the basis of cost predictions, to overall cost estimations of different construction projects including infrastructure and residential buildings.

ANNs for productivity modelling and estimation

Another vital task in the construction management field is to provide productivity rates and measurement, especially in areas such as resource allocation and management, scheduling, estimating, accounting, cost control and payroll (Herbsman and Ellis 1990). Contractors should have reliable estimates for different construction operations which are needed for planning and scheduling purposes. ANNs can be utilized in this area as a robust tool for productivity modelling and estimation. Some of the construction processes and operations for which ANNs have been used to model productivity are:

- concrete pouring, formwork and finishing tasks (Portas and AbouRizk 1997; Ezeldin and Sharara 2006; Dikmen and Sonmez 2011)
- hoisting times (hook times) of tower cranes (Tam *et al.* 2004)
- pile construction (Zayed and Halpin 2005)
- plastering activities (Oral *et al.* 2012)
- earth-moving equipment (Chanda and Gardiner 2010; Hola and Schabowicz 2010; Han *et al.* 2011).

Most of these studies show the application of ANNs for productivity modelling at the task level. ANNs are, however, applicable to any level of productivity prediction just as they can be applied to various levels of cost estimation.

Based on the previous studies, Figure 11.3 depicts a framework that shows all of the steps required to develop efficient ANNs for productivity and cost estimation.

Example: estimating on-site productivity of precast installation activities

Problem formulation

A contractor in charge of the installation of different precast elements wanted to have more accurate estimates of on-site erection activities in order to manage the required resources more efficiently. An estimator was

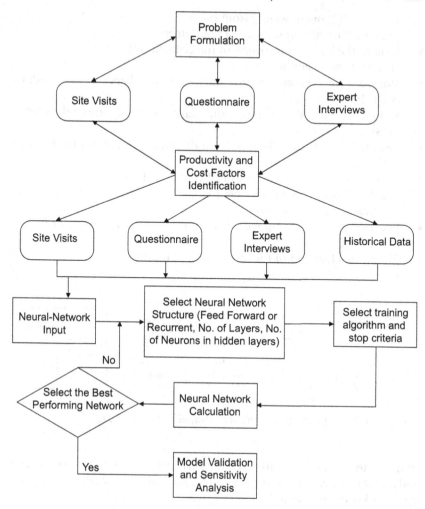

Figure 11.3 A framework for the development of an ANN for productivity and cost estimation

hired to provide a tool which could predict the time needed to install different precast components such as walls, columns, beams and slabs. For better understanding of the problem requirements, several preliminary site visits were conducted, and the estimator determined that a typical erection process or cycle includes several activities and resources; these are shown graphically in Figure 11.4.

Factors affecting productivity of precast installation

Based on Figure 11.4 and through site visits, expert interviews and a literature review, the following factors were identified as being important for productivity of precast erection processes:

- Weight: component weight (tonnes)
- Area: the largest surface area of the element (m²)
- Length: the longest dimension of the element (m)
- Height: component height (m)
- Storage type: the component is stored among others or isolated (0: isolated; 1: among others)
- Storage to crane: distance from the element to the centre of the crane at the storage area (m)
- Installation to crane: distance from the installation point to the centre of the crane (m)
- Crane angle: angular movement of crane (degrees)
- Installation type: the component is installed among others or isolated (0: isolated; 1: among others)
- Location type: the component is part of the exterior or interior (0: interior; 1: exterior)
- Elevation: elevation of the installation point (m).

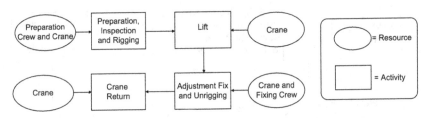

Figure 11.4 Precast installation activities (Najafi and Tiong 2012)

Other factors such as weather conditions, management conditions, crew skill and the like could also be considered; however, for simplicity, they were not included in this example.

Data collection

The next step was to collect the required data needed for ANN training and validation. In this case, the estimator collected the data regarding the installation of 91 precast elements (walls, columns and slabs) through site observations. General specifications of the collected data are shown in Table 11.1.

Table 11.1 shows that the observed precast components are among the typical elements that are widely used in construction projects. The results of the study can therefore be generalized and applied to new projects as well.

Selecting the ANN architecture and training algorithm

The estimator then used a commercial package (e.g. NeuroSolutions™) to design the ANN. For simplicity, a typical feed-forward back-propagation

Table 11.1 General specifications of the collected data

Type	Number of cases	Length (range in metres)	Width (range in metres)	Height (range in metres)	Weight (range in tonnes)
Wall	39	1.40–5.75	0.10–0.25	2.80	0.95–7.31
Column	28	0.80–2.00	0.20–0.25	2.80	1.00–2.40
Slab	24	2.40–5.34	1.00–2.40	0.07	0.80–2.10

network was chosen as the main architecture. The inputs of the network were the factors described in the previous sections, and the main output is the total installation time in minutes.

From 91 observations, 60 data points were randomly selected for the training of the network, 16 cases for cross-validation, and the remaining 15 data points were used for validation purposes. Use of the cross-validation technique ensures that the network is not overtrained (a situation in which the network performs well on the training set but performs poorly on the test data). With this method, the stop criterion is chosen in such a way that as soon as the error in the cross-validation set starts to increase, the training will be stopped. Since there is no specific rule for the configuration of ANNs, the estimator should develop several ANNs and choose the one with minimum error.

Table 11.2 shows different ANN architectures and the corresponding mean squared error (MSE) and mean absolute error (MAE), which are calculated using the following formulae:

$$MAE = \frac{\sum_i^n |Y_{NN} - Y_A|}{n}$$

(11.1)

$$MSE = \frac{\sum_i^n (Y_{NN} - Y_A)^2}{n}$$

(11.2)

Y_{NN} is the network output and Y_A is the actual (desired) output for each testing data point. The best-performing network, the one that contains the minimum error (min. MSE), is highlighted. Note that in Table 11.2, architecture 11–7–1 denotes a network with three layers (one hidden layer) containing 11, 7 and 1 neurons in the input, hidden, and output layers, respectively.

Validation and performance of the selected model

A set of 15 data points was used to test the performance of the model against the actual data collected from the construction sites, and the results are shown in Figure 11.5.

Table 11.2 Selection of the best performing network

Architecture	11–7–1	11–9–1	11–11–1	11–13–1	11–15–1	11–17–1
MSE	7.855	7.712	7.581	7.273	6.233	6.416
MAE	2.405	2.345	2.327	2.300	2.033	2.110

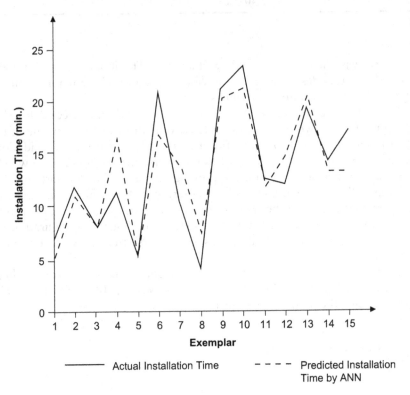

Figure 11.5 Predicted vs. actual installation times

The mean absolute percentage error (MAPE) calculated by Equation 11.3 was equal to 19.86%, which means that the installation times predicted by the ANN are about 20% higher or lower than the actual installation times, which can be considered to be an acceptable performance considering the uncertainties involved in construction estimations.

$$MAPE = \left(\sum_{i}^{n} 100 * \frac{|Y_{NN} - Y_A|}{Y_A} / n \right)$$
(11.3)

Similar steps can be used to develop ANNs for productivity estimations of other construction operations or cost predictions for different construction projects. The main differences are in the factors affecting productivity or cost estimations and the procedure followed to collect the required data. It should be noted that the aforementioned errors can be further minimized by increasing the sample size of the collected data (training set). Additionally, the use of other network types with different learning algorithms may result in better estimates. Statistical tests and analysis are also recommended for further comparison between the characteristics of the actual data and the estimated data provided by ANNs.

Hybrid use of artificial intelligence

The performance of ANNs can be further enhanced when they are used together with other artificial intelligence techniques such as fuzzy logic and genetic algorithms. Basically, fuzzy and neural systems are structurally different; however, they share a complementary nature and can be integrated to improve the overall flexibility and expressiveness of neural networks (Tsoukalas and Uhrig 1997). ANNs can be applied in function approximation, pattern recognition and automatic learning. On the other hand, fuzzy logic is efficient in modelling uncertainties. Their combination is therefore useful in dealing with certain construction management research problems that require accurate approximation while including different types of uncertainties. The performance of 'neurofuzzy' systems can be further improved by using genetic algorithms (Holland 1992) to optimize ANN parameters such as connection weights or network topologies (number of neurons in the hidden layer).

Additionally, with the increasing adoption of building information modelling (BIM) packages, there is a need to develop ANN analysis as a separate module that can be integrated with BIM systems. Architects, engineers, contractors and owners use BIM programs to plan, collaborate and predict performance before breaking ground on buildings. This chapter shows that ANNs are able to provide accurate productivity modelling of any construction operation, and the integration of ANNs with BIM systems is beneficial for both tools. This can be shown by the example of precast installation. Note that to use ANNs to estimate the installation time of each precast element, an estimator should manually feed all of the required input data such as element weight, surface area, the longest length and so on, as these are the factors affecting productivity in precast installation. Since most of this data is already available in the BIM model of the building, a computer program can be developed to automatically extract the required input for ANNs from the BIM system and perform the analysis. The results from ANNs can be then linked back to the BIM system to predict the performance more accurately.

Conclusions

Artificial neural networks are an information-processing technology that simulates the human brain system through learning from different experiences and generalizing from previous examples. Since the 1990s, this technique has been widely used in various areas of construction engineering and management, including cost and project duration estimations, bidding models and mark-up estimations, productivity modelling, prequalification of contractors and risk management.

This chapter summarized how ANNs can be easily applied to the two most important construction management tasks, cost estimation and productivity modelling. The procedures for using ANNs in other areas are similar to those depicted in the framework described here.

Some construction management problems deal with different levels of uncertainty. In these cases, the use of 'neurofuzzy' systems is highly recommended to add the flexibility required to solve these problems with the integration of ANNs with fuzzy systems. Additionally, by increasing adoption of BIM packages, the integration of BIM systems with ANNs is suggested in order to automate the data entry process as well as to improve the accuracy of performance prediction in the BIM system.

Finally, it is recommended that construction management research scholars and industry players further investigate the applications of ANNs. There are several commercial and open-source packages that provide user-friendly interfaces which can be easily used to develop efficient ANNs with only basic knowledge of this technique. Although ANNs have been used for more than two decades in the construction industry, further research gaps can be identified through extensive literature review, especially in those areas where ANNs have been utilized in other industries.

References and Further Reading

Alavala, C. (2006). *Fuzzy logic and neural networks: basic concepts and applications*. New Age International, New Delhi.

Attalla, M. and Hegazy, T. (2003). "Predicting cost deviation in reconstruction projects: artificial neural networks versus regression." *Journal of Construction Engineering and Management* 129 (4), 405–411.

Baalousha, Y. and Çelik, T. (2011). "An integrated web-based data warehouse and artificial neural networks system for unit price analysis with inflation adjustment." *Journal of Civil Engineering and Management* 17 (2), 157–167.

Bhoka, S. and Ogunlana, S. (1999). "Application of artificial neural network to forecast construction duration of buildings at the predesign stage." *Engineering, Construction and Architectural Management* 6 (2), 133–144.

Boussabaine, A. H. (1996). "The use of artificial neural networks in construction management: a review." *Construction Management and Economics* 14 (5), 427–436.

Chanda, E. and Gardiner, S. (2010). "A comparative study of truck cycle time prediction methods in open-pit mining." *Engineering, Construction and Architectural Management* 17 (5), 446–460.

Chen, W. T. and Huang, Y. H. (2006). "Approximately predicting the cost and duration of school reconstruction projects in Taiwan." *Construction Management and Economics* 24 (12), 1231–1239.

Dikmen, S. U. and Sonmez, M. (2011). "An artificial neural networks model for the estimation of formwork labour." *Journal of Civil Engineering and Management* 17 (3), 340–347.

Emsley, M. W., Lowe, D. J., Duff, A. R., Harding, A. and Hickson, A. (2002). "Data modelling and the application of a neural network approach to the prediction of total construction costs." *Construction Management and Economics* 20 (6), 465–472.

Ezeldin, A. S. and Sharara, L. M. (2006). "Neural networks for estimating the productivity of concreting activities." *Journal of Construction Engineering and Management* 132 (6), 650–656.

Günaydın, H. M. and Doğan, S. Z. (2004). "A neural network approach for early cost estimation of structural systems of buildings." *International Journal of Project Management* 22 (7), 595–602.

Han, S., Hong, T., Kim, G. and Lee, S. (2011). "Technical comparisons of simulation-based productivity prediction methodologies by means of estimation tools focusing on conventional earthmovings." *Journal of Civil Engineering and Management* 17 (2), 265–277.

Haykin, S. S. (1999). *Neural networks: a comprehensive foundation.* Upper Saddle River, NJ: Prentice Hall.

Hegazy, T. and Ayed, A. (1998). "Neural network model for parametric cost estimation of highway projects." *Journal of Construction Engineering and Management* 124 (3), 210–218.

Herbsman, Z. and Ellis, R. (1990). "Research of factors influencing construction productivity." *Construction Management and Economics* 8 (1), 49–61.

Hola, B. and Schabowicz, K. (2010). "Estimation of earthworks execution time cost by means of artificial neural networks." *Automation in Construction* 19 (5), 570–579.

Holland, J. H. (1992). "Genetic algorithms." *Scientific American* 267 (1), 66–72.

Kim, G., Yoon, J., An, S., Cho, H. and Kang, K. (2004). "Neural network model incorporating a genetic algorithm in estimating construction costs." *Building and Environment* 39 (11), 1333–1340.

Najafi, A. and Tiong, R. (2012). "A study of productivity of precast concrete installation." *Management in Construction Research Association (MiCRA) Postgraduate Conference.* University Teknologi Malaysia (UTM), Kuala Lumpur, Malaysia.

Oral, M., Oral, E. and Aydin, A. (2012). "Supervised vs. unsupervised learning for construction crew productivity prediction." *Automation in Construction* 22, 271–276.

Petroutsatou, K., Georgopoulos, E., Lambropoulos, S. and Pantouvakis, J. (2011). "Early cost estimating of road tunnel construction using neural networks." *Journal of Construction Engineering and Management* 1 (1), 351.

Portas, J. and AbouRizk, S. (1997). "Neural network model for estimating construction productivity." *Journal of Construction Engineering and Management* 123 (4), 399–410.

Rumelhart, D., Hintont, G. and Williams, R. (1986). "Learning representations by back-propagating errors." *Nature* 323 (6088), 533–536.

Tam, C., Tong, T. and Tse, S. (2004). "Modelling hook times of mobile cranes using artificial neural networks." *Construction Management and Economics* **22** (8), 839–849.

Tsoukalas, L.H. and Uhrig R.E. (1997). *Fuzzy and neural approaches in engineering.* John Wiley & Sons, Inc., New York, NY.

Wichan, P., Thammasak, R. and Vanee, S. (2009). "Forecasting final budget and duration of highway construction projects." *Engineering, Construction and Architectural Management* **16** (6), 544–557.

Williams, R.W. and Herrup, K. (1988). "The control of neuron number." *Annual Review of Neuroscience* **11**: 423–453.

Wilmot, C.G. and Mei, B. (2005). "Neural network modeling of highway construction costs." *Journal of Construction Engineering and Management* **131** (7), 765–771.

Zayed, T.M. and Halpin D.W. (2005). "Pile construction productivity assessment." *Journal of construction engineering and management* **131** (6), 705.

Editorial comment

We regularly see and hear advertisements for laundry detergent that claim that a product has 'twice the cleaning power of other detergents' or is 'guaranteed to make your coloured clothes 30% brighter'. Such claims are dubious at best, and no mention is made of how these performance measurements might be made. There are other comparisons that we seek to make where finding a way to measure performance is equally difficult. For example, how might we assess the relative performance or efficiency of a number of branches of the same bank? Some possibilities come to mind: maybe it could be the number of transactions completed per week or the amount of money deposited and withdrawn in a period of time. Closer consideration of these possible methods quickly reveals the flaws: for example, a couple of very large transactions could have a high dollar figure yet not indicate in any way that the branch was running efficiently.

Another variable that could be considered is the number of staff employed at each branch, as a greater number of transactions might be processed in one branch but be utilizing more staff in order to do that. It is in situations like this that the technique of data envelopment analysis (DEA) can be used. If measuring productivity is the aim, then the typical measure of productivity, output over input, is still valid, but it is necessary to identify what will represent output and what will represent input. In the bank branch example, the number of transactions might represent output, with input being the number of staff. Taking the ratio of output to input for different branches will give some indication of relative efficiency – it is important to note that it is *relative* efficiency, not any absolute measure of efficiency that is obtained. That, however, is enough to help management identify which branches are performing better than others, where fewer staff might be deployed and so on.

In the following chapter, the application of DEA to the measurement or assessment of construction productivity is reviewed. It provides another approach to the benchmarking of construction industry performance (see Chapter 8) and also provides a way to assess changes in productivity over time. While DEA has been used in a number of studies related to construction over several decades, there is considerable scope for further development and application of this technique.

12 Determining levels of productivity in the construction industry using data envelopment analysis

Malcolm Abbott

Introduction

Over the years, numerous attempts have been made to determine the main drivers of productivity and efficiency in the construction industry. Most studies that look at the productivity of the construction industry are centred on the industry in the United States, although in more recent years, studies have been undertaken in of other countries as well. In the United Kingdom, for instance, data for a range of performance indicators in the industry is now collected by the national government (BERR 2008).

One element some researchers historically found was that there was a perceived stagnation or decline in productivity growth in the United States construction industry, and these researchers had some difficulty explaining why this might have occurred. This work includes research by Stokes (1981), Allen (1985) and Schriver and Bowlby (1985).

One influential view expressed was that in the past, the construction industry had been largely a labour-intensive one, which meant that the introduction of new equipment and materials could only increase output with a given level of labour and capital. Another view was that the stagnation could be explained by the use of inappropriate price indices in determining productivity measures (Dacy 1965; Stokes 1981) or by the fact that the official data on construction industry output did not account for improvements in the quality of output over time (Rosefielde and Mills 1979). A final view was that much of the stagnation in productivity could be attributed to a change in the output mix from high- to low-productivity building projects (Schriver and Bowlby 1985).

Whatever the reason, it seems that determining levels of productivity in the industry involves a number of difficulties. Overcoming these difficulties has meant that some attempts have been made to develop and use a range of different productivity analysis tools to determine the level of productivity in the construction industry and to identify some of the main drivers of productivity growth. One of the measures that has begun to be used is the method known as data envelopment analysis (DEA). DEA is a technique that has been around for a number of years and has generally been used as a means of benchmarking the performance of different companies against one another in a range of different industries. In addition, it has been used to determine changes in the productivity of industries over time (effectively benchmarking an industry against itself in different years).

In this chapter, therefore, a description of the manner in which DEA is used in general and how it has been used in the case of the construction industry is provided. In doing so, examples of pure benchmarking of companies and construction projects are provided as well as some examples of productivity measures and how they change over time. In undertaking this, some suggestions for future research are provided.

Data envelopment analysis

Companies and government service providers often operate in markets which lack prices and costs determined under competitive conditions. In these cases, the usual market indicators of performance like profitability and rates of return cannot be used to gauge a firm or industry's economic performance accurately (often they instead indicate a company or government agency's degree of market power). In these cases, it is more appropriate to use productivity measures that do not rely on output and input prices. DEA is a technique that achieves this. In the case of the construction industry, it has been argued that construction materials and property price indices do not accurately reflect the input costs and output prices of the industry.

DEA is a linear programming technique which estimates organizational efficiency by measuring the ratio of total inputs employed to total output produced for each organization. This ratio is then compared to others in a sample group to derive an estimate of relative efficiency. Organizations in the sample group receive an efficiency score determined by the variance in their ratio of inputs employed to outputs produced relative to the most efficient producer in the sample group. The advantage is that it can be used without input or output prices, as simple volume figures can be used to indicate inputs and outputs. DEA has the advantage of being a nonparametric technique, and it avoids the need to make assumptions regarding the functional form of the best-practice frontier (e.g. Cobb-Douglas or trans-log). DEA can also incorporate multiple inputs and outputs and can be used to calculate technical and scale efficiency with information on output and input volumes.

DEA was pioneered by Charnes, Cooper and Rhodes (1978) and was based on the work by Farrell (1957) (see, for example, Knox Lovell and Schmidt 1988; Färe, Grosskopf and Knox Lovell 1985 and Coelli, Rao and Battese 1998). DEA estimates technical efficiency by finding the difference between the observed ratios of output to input and the ratios achieved by the best-practice decision-making units (DMUs). It can be expressed as the potential to increase the quantity of outputs from given quantities of inputs or the potential to reduce the quantities of inputs used in producing given quantities of output. Technical efficiency is affected by factors such as the size of the operation, managerial practices, ownership structure and the regulatory environment. The concept can be depicted as shown in Figure 12.1, which plots different combinations of two input-to-output ratios – labour to output and capital to output. – required to produce a given output. This curve is known as the *unit isoquant* or *best-practice frontier*. If a DMU is producing on the unit isoquant (for example at points A or B), then it is regarded as technically efficient. DMUs operating to the right of the unit isoquant (for example at point C) are considered to be operating with a higher degree of technical efficiency. Knowledge of factor prices would allow identification of which DMU, for example A or B in Figure 12.1, is allocatively efficient – that is, it uses the appropriate mix of inputs.

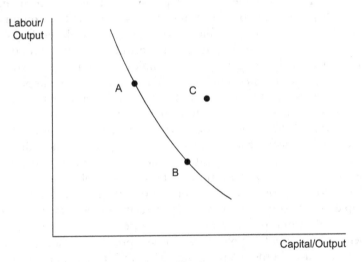

Figure 12.1 Illustration of the technical efficiency concept

Scale efficiency is the extent to which a DMU can take advantage of returns to scale by altering its size towards the optimal scale (which is defined as the region in which there are constant returns to scale in the relationship between outputs and inputs). Figure 12.2 shows production frontiers under both constant and variable returns to scale in the case of a single input and single output. The constant returns to scale production frontier is the scale line from the origin (O-X). The variable returns to scale frontier (VRS) passes through the points where the DMUs have the highest output to input ratios given their relative sizes. DEA has both an input and output orientation, and VRS scores will vary depending on the orientation. The scale efficiency of each DMU can be determined by comparing the technical efficiency scores of each DMU under constant returns to scale and variable returns to scale.

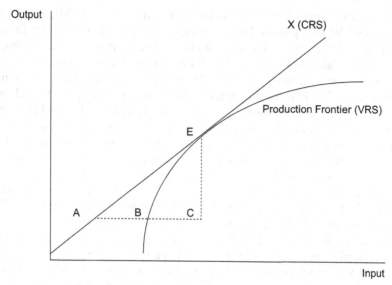

Figure 12.2 The production possibilities frontier and returns to scale

It is important to separate technical and scale efficiency, as they provide different pieces of information. For example, DMU C (in Figure 12.2) is technically inefficient as measured by the distance between B and C. DMU C could produce the same volume (and quality) of output with less input (move from C to E). DMU B is technically efficient as it lies on the VRS frontier and it is scale efficient (measured by the distance A to B). The distance from the respective frontiers determines technical efficiency under each assumption. The distance between the constant returns and variable returns frontiers determines the scale efficiency component. Technical efficiency resulting from factors other than scale is determined by the distance from the variable returns frontier. Thus when efficiency is assessed under the assumption of variable returns, the efficiency scores for each organization indicate only technical efficiency resulting from nonscale factors such as an economical use of resources and competent management.

Productivity change, on the other hand, is a dynamic indicator of the change in outputs relative to inputs over time. Productivity growth will reflect changes in technical and allocative efficiency, technological improvements and changes in the external environment in which production occurs. A firm improves its productivity either by moving towards a best-practice frontier or when the frontier shifts outwards due to technological change. The Malmquist DEA approach uses panel data to estimate changes in technical efficiency, technological progress and total factor productivity. The Malmquist DEA approach has been used in the past in a variety of circumstances such as for financial institutions (Worthington 1999; Berg,

Forsund and Jansen 1992), electricity utilities (Färe *et al.* 1990), gas utilities (Price and Weyman-Jones 1996), hospitals (Färe *et al.* 1994) and airports (Abbott and Wu 2002). These methods are discussed in Färe *et al* 1994. In effect, the Malmquist DEA approach derives an efficiency measure for one year relative to the prior year while allowing the best practice frontier to shift.

Figure 12.3 illustrates this point in terms of a one-output (Y) and one-input (X) model. The production frontier in period 1 is F1, and in period two it is F2. Assume that in period 1 the firm employed X1 units of the input, and if it was technically efficient, it would have produced y1* units of output. If only Y1 units of output were produced in period 1, then the firm is technically inefficient. The vertical distance y1*to y1 is a measure of technical inefficiency. Assume that in period 2, the firm employed X2 units of the input, and if it was technically efficient, it would have produced y2*'' units of output. If output in period 2 is given by y2, then the firm is technically inefficient (measured by the vertical distance y2*'' to y2. The change in technical inefficiency over the two time periods is given by the difference between y1* to y1 and y2*'' to y2. Technological progress is measured by the displacement of the frontiers. This can be done by comparing y1* and y1*'' (in terms of input X1) or y2* to y2*'' (in terms of input X2). So, when comparing the actual points of y1 in period 1 and y2 in period 2, we can decompose the actual output growth (y1) into input growth, measured by X2 – X1; technological progress, measured by say y2* to y2*'', and technical efficiency change, measured by the difference between y1* to y1 and y2*'' to y2.

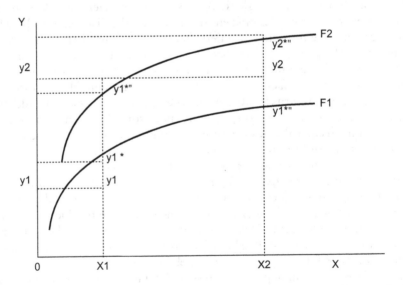

Figure 12.3 Decomposition of output growth into input growth, technical change and efficiency change

In the case of the construction industry, it would appear that it would be possible, if sufficient data is available, to use DEA either as a benchmarking tool to show the relative efficiency of DMUs at any point in time or to determine the change in productivity over time.

Productivity change over time

Although DEA was originally devised as a benchmarking tool to show the relative performance of DMUs against each other, the most common use of it in the construction industry has been to determine changes in productivity over time at the industry level (see Table 12.1 for a number of examples). To carry out this estimation, the DEA Malmquist approach has been used in a number of cases to determine the manner in which productivity has changed over time. In doing so, the technique was used in the manner described in the previous section.

The methods used have tended to make use of construction data at the national level, although two of the studies have made used of provincial- or state-level data to attempt to increase the size of the data sample (see Xue *et al.* 2008; Li and Liu 2011). Industry-level studies of this sort are of some interest to policy makers, as they describe the state of an industry's development in general but are perhaps of limited interest to those managing units within the industry.

Probably the first example of the DEA Malmquist approach being used to determine the change in productivity in the construction industry over time was the study undertaken by Färe, Grosskopf and Margaritis (1996) for the New Zealand construction industry. This work was undertaken as part of a broad economy-wide study of productivity change in New Zealand, where the work on the construction industry was just one of a number of industry-level studies undertaken.

Färe and Grosskopf were two of the pioneers of the use of DEA more generally, and they teamed up with the New Zealand economist Margaritis to undertake this work. They matched the value added of the construction industry (output) to the inputs of net capital and labour hours employed in the industry to determine the productivity growth in the industry in New Zealand between 1973 and 1994 (a similar approach was used in analysing the other industry sectors of the New Zealand economy). The study determined technical and scale efficiency levels and their change over time, along with the impact of technological change. They did not make use of price indices and so did not determine changes in the levels of allocative efficiency.

The results of this first study showed an average –0.4 per cent (i.e. negative) growth rate over the period, which was in line with other studies of the industry over the period (mostly done using index approaches). The period 1974 to 1994 was a rather chequered one in New Zealand's economic history, containing as it did three recessions (1974, 1981, 1991), along with a period of high inflation in the industry and a serious loss of markets when

Table 12.1 Past studies on the productivity of the construction industry using DEA

AUTHOR(S)	DATE	JURISDICTION	PERIOD STUDIED	VARIABLES and RESULTS
Benchmarking exercises				
Edvardsen	2005	Norway	2001	342 companies Output: sales values. Inputs: labour man-years, rental expenditure and depreciation Results: divergence of companies, most efficient are those with high wages, long hours of work, few apprentices, diversified output.
Ingvaldsen	2005	Norway	2003 to 2005	138 building projects (block of flats) Outputs: square metres. Inputs: labour hours, materials, energy, rent of plant and equipment Results: divergence in efficiency between projects
McCabe, Tran and Ramani	2005	Canada	1998 to 2000	Ten contracts over three years. Outputs: sales history, employee experience. Inputs: safety record, capacity, related experience Found differences in relative efficiency scores
El-Mashaleh, Minchin and O'Brien	2007	United States (Florida)		74 firms Output: schedule performance, cost performance, customer satisfaction, profit. Inputs: management expenses, safety expenses Results: found differences in relative efficiency scores
Chiu and Wang	2011	Taiwan	1999 to 2008	27 companies over a 10-year period Inputs: Stockholders' equity ratio, turnover rate of inventory, turnover rate of accounts receivable Outputs: return of stockholders' equity, operating income ration, quick ratio, cash flow ratio, return of assets Results: demonstrable difference in the financial efficiency of the firms
Horta, Camanho and Moreira da Costa	2012	Portugal	1996 to 2009	110 companies Inputs and outputs: Liquidity, autonomy, value added, profitability, Steady improvement in productivity over time

DEA Malmquist

Färe, Grosskopf and Margaritis	1996	New Zealand	1973 to 1994	Industry wide. Output: value added. Inputs: labour hours, net capital. Results: productivity growth of average-0.4% per annum
Wang	1998	Hong Kong	1981 to 1994	Industry wide. Output: gross output of contractors. Input: labour employed, materials expenditure, value of fixed assets. Productivity growth: 2.9% per annum
Wang and Chau	1997	Hong Kong	1981 to 1994	Industry wide. Output: value added. Input: labour employed rents/rates/interest/depreciation on capital. Results: productivity growth of average 2.6% per annum
Wang and Chau	2001	Hong Kong	1981 to 1996	Industry wide. Output: value added. Input: labour employed, rents/rates/interest/depreciation on capital. Results: increasing trend of productivity over the period
Chau and Wang	2005	Hong Kong	1981 to 2002	Industry wide. Output: gross output of contractors. Input: labour employed, materials expenditure, value of fixed assets. Results: productivity growth of average 2.6% per annum.
Choy	2008	Malaysia	1970 to 2004	Industry wide. Output: gross output. Input: employees, value of assets. Results: increasing trend of productivity over the period
Xue, Shen, Wang and Lu	2008	China	1997 to 2003	Provincial level. Output: value added. Inputs: employees, asset values. Results: improvement in productivity, especially in eastern provinces
Wang, Ye and Yuan	2010	China	1997 to 2007	Industry wide. Outputs: value, profit, waste. Inputs: employees, total assets, other costs. Results: significant growth in productivity
Li and Liu	2011	Australia	1990 to 2008	State based. Output: value added. Input: employees, fixed assets. Results: improved productivity at an average of 0.629% per annum
Chiang, Li, Choi and Man	2012	China and Hong Kong	2004 to 2010	20 construction companies. Inputs: value of assets, employees, cost of materials. Outputs: Revenue, profit. Results: Convergence in efficiency of Chinese to Hong Kong levels

Britain joined the European Economic Union. In the case of the New Zealand construction industry, demand stagnated (due to substantial emigration to Australia), and investment in the industry fell. Productivity growth in the construction industry in the period, therefore, is usually regarded as being fairly slow due to a combination of low capacity utilization and low investment in new technologies.

Since this work by Färe, Grosskopf and Margaritis was first undertaken, other studies have been completed in a number of other countries. Examples include work by Chau and Wang (2005), Wang (1998) and Wang and Chau (1997, 2001) on the construction industry in Hong Kong, Choy (2008) on the Malaysian industry, Wang et al. (2010) and Xue et al. (2008) on the Chinese industry and Li and Liu (2011) on the Australian industry (summaries of these studies are provided in Table 12.1).

Unlike the New Zealand study, in each of these later cases, the researchers found considerable increases in productivity over the studied periods. This was perhaps not unexpected, as in each case (except the Australian one), the studies were of developing countries in which the employment of additional capital drove productivity growth. In each case, investment in capital accumulation and the introduction of technological developments in the construction industry were high. Rapid technological change was, therefore, found to be the experience in each case, and these are shown in the results of each of these studies. In these cases, what occurred was that over time, the best-practice frontier moved outwards due to the introduction of more advanced technologies.

In terms of the selection of the outputs and inputs in each of these studies, the inputs tended to be concentrated on combinations of labour (either work hours or the number of employees) and estimations of the value of capital. In some cases, other inputs such as materials were also included (see, for instance, Wang 1998; Chau and Wang 2005). In the case of the value of capital inputs, in some cases a simple valuation of fixed assets was used (Xue et al. 2008; Li and Lu 2011) or, alternatively, an estimation of the cost of capital in terms of rents, interest and depreciation (Wang and Chau 1997). Output tended to be determined by some valuation of the output of the industry, based on expenditure or the value added of construction from official statistical sources. No attempt has been made in any case to use other indicators of output such as square metres of floor space.

In each case, this approach ran the risk of the expenditure figures not capturing all of the improvements in the quality of the materials and design of construction projects or a change in the composition of the industry – residential versus commercial or engineering projects – or indeed other intangibles in the industry such as the improvement of safety standards or times for completion of projects.

It was probably the case in each study that the approach was chosen because of the ease of collecting the necessary data from official sources. There is nothing especially wrong with this, although it does ignore the issues of quality that were originally raised back in the 1970s and 1980s. It does mean, however, that there is only so much that can be learnt from these studies in terms of identifying the main drivers of growth in the construction industry.

The results of the studies should, therefore, be seen as crude estimations of changes in industry-level productivity, results that might be open to modification and reinterpretation. In addition, what they did was to identify some of the broad drivers of productivity change over time (such as improvements in technical and scale efficiency and technological change), but they did not dig down any deeper to identify what some of the more specific drivers of productivity growth were. A range of factors can, for instance, contribute to raising the technical efficiency of firms and industry. Better management techniques, higher levels of skill or greater external economies can all lead to improved levels of technical efficiency. The methods used and the data collected in each case do not allow for a more detailed understanding of the main drivers of efficiency in the construction industry.

It might be the case that this approach, using very highly aggregated data, cannot really determine what the drivers of productivity change in the construction industry are. To do that, perhaps DEA would be better used in conjunction with more data-intensive firm- or project-level studies. This approach, however, is dependent to a substantial degree on the ability to obtain very detailed data at the project or firm level. It is probably not surprising, therefore, that there have been few studies of this sort undertaken.

Benchmarking

Efficiency studies using DEA to carry out benchmarking exercises of the construction industry at the firm or project level are few and far between, even though it is possible that this approach would be a useful one. There have been some attempts in the past (see Table 12.1 for some examples) to use DEA in that fashion and with some effect, and it is likely that this approach will be increasingly used.

Some examples of DEA benchmarking in the construction industry include Edvardsen (2005) on Norwegian construction firms, Ingvaldsen (2005) on Norwegian building projects, McCabe, Tran and Ramani (2005) on Canadian construction firms, El-Mashaleh, Minchin and O'Brien (2007) on firms in Florida in the United States and Chiu and Wang (2011) on firms in Taiwan.

Problems with using DEA in benchmarking firms and projects in the construction industry are twofold. First of all, a large enough sample of data is required from individual firms, which in many cases is not easily obtained; firms generally are quite reluctant to share information about their commercial operations. The second problem involves the comparability of firms and projects. In the construction industry, no single construction project is exactly the same as any other, and no construction company is, strictly speaking, the same in composition as any other. For that reason, a number of the studies in this area fall back on using aggregated figures of value added as indicators of firm output. It also explains why in one study it was decided to compare and benchmark the efficiency of projects that were of a broadly similar type, being three-storey blocks of flats (Ingvaldsen 2005).

It is notable, however, that the choice of inputs and outputs when using DEA as a benchmarking tool is more varied than in the case of DEA Malmquist studies at the industry level. Two past cases, for instance, used a range of financial ratios to determine the efficiency of firms: Chiu and Wang (2011), who looked at Taiwanese construction firms, and Horta, Camanho and Moreira da Costa (2012), who looked at Portuguese construction companies. Ingvaldsen (2005) used floor space as an indicator of output for construction projects in Norway.

Despite these more varied approaches and more extensive use of different data, most of the studies came up with results that are unremarkable. That is, they found that a range of companies and projects have quite different levels of efficiency, which in turn means that there is scope for some companies, at least, to improve their levels of efficiency and move towards a best-practice frontier. In terms of explaining why some firms are below the best-practice level efficiency, the studies find that it is either because of a lack of technical efficiency or scale efficiency. These sorts of conclusions do not go very far in explaining why some operations might be at suboptimal levels of efficiency.

In the context of past work using DEA in other industries, these sorts of results are not especially surprising. Generally, DEA is first used to determine if the DMUs in a given sample show some signs of inefficiency compared to best-practice levels. Later work then starts to try to identify specifically why these inefficiencies exist and determines what can be done about them. To date, not enough studies have been undertaken using DEA to provide the impetus for much movement towards the later stage.

Two papers on the construction industry do, however, go further, and they give a reasonable indication of how this might be achieved. The most important of these was by Edvardsen (2005) who looked at the relative efficiency of a range of Norwegian construction companies. Although Edvardsen followed an approach that is not at all common in the world of construction DEA, it is a fairly common approach in the case of DEA studies of firms in other industries.

What Edvardsen did was to determine the efficiency levels for 342 construction companies in Norway in a single year using DEA by showing the relationship between the sales of the firms as the output and indicators of labour and capital as the inputs (labour man-years and rental expenditure and depreciation). Instead of leaving it at that point, as other studies have tended to, he then performed a second-stage regression that ran the relative efficiency levels of the firms as the dependent variable and a range of other variables as independent variables in order to identify if the latter variables had any impact on the level of efficiency in the industry.

The independent variables he used were such things as high average wages, long average hours of work, the number of apprentices employed, how diversified output was between different types of construction activity and the location of the main work of the companies, either in the main Norwegian city of Oslo or elsewhere. The purpose in determining the relationship among these variables and the level of efficiency in the firms was to determine if they might have made some contribution to the relative efficiency levels.

Edvardsen found that there was a correlation with having highly skilled, well-paid workers who worked long hours on the one hand and firms' levels of efficiency on the other. It also found that the more diversified companies had higher levels of efficiency. Companies located mainly in Oslo were not found to have any significant statistical relationship with higher levels of efficiency. This does not necessarily mean that there is no relationship between efficiency and the ability of firms to operate in a large urban centre, but this study was statistically unable to detect if it was so or not.

In the paper by Horta, Camanho and Moreira da Costa (2012), a second-stage regression was also undertaken using the variables of the size of GDP, the size of the firms, R&D engagement and whether the firm was located in the major urban centres of Lisbon. The growth of GDP and the size of the firms were both positively correlated with levels of efficiency.

The independent variables chosen in these two papers are by no means the only ones that could have been used to identify the main drivers of efficiency. Such things as the safety standards of companies, regulatory impediments to construction, the size and scope of companies, client satisfaction with construction work, the qualification levels of the workforce and the timely completion of projects are all variables that might have been used if the data had been available. Many of these variables are collected by statistical and other government agencies and so might be used in further work to more specifically identify the main drivers of efficiency in the industry.

In the case of project-level benchmarking, it might be possible to take similar projects and use DEA to benchmark them against each other and then undertake a second-stage regression in order to identify important correlations. By doing so, it might be possible to understand better why there is often a divergence in the level of efficiency between projects that is achieved across a sample of firms in the industry.

It would seem that there is still considerable scope for further work of this sort in the future in relation to the construction industry. One advantage of using DEA in the case of the construction industry is that it is an industry that often has a number of firms that can be used to benchmark one another. Despite the difficulties that exist, it would be expected that further attempts will be made to use DEA in a more adventurous fashion. It might be that this approach would tell us more about the main causes of efficiency (and inefficiency) in the industry than the more often used approach of industry-level DEA Malmquist productivity indexes.

Conclusion

Although DEA has been around for a number of years, its use in the benchmarking of firms and projects in the construction industry has been limited to date. This is unfortunate, as in some ways it seems an ideal way to determine what the causes of efficiency and inefficiency are in the industry.

One persistent problem has been a lack of firm-level data that could be used by researchers. This is probably the main reason why researchers have in the past had a tendency to undertake studies at an industry level, using official sources of data, over time.

These very aggregated studies can tell us something in general about the course of productivity change in the industry over time but do not provide very detailed information about the causes of efficiency and inefficiency in the industry.

In future, it is hoped that more detailed benchmarking studies will be able to be undertaken that will assist in developing a deeper understanding of the nature of the construction industry.

References and Further Reading

Abbott, M. and Wu, S. (2002) 'Total factor productivity and efficiency of Australian airports', *Australian Economic Review*, 35 (3), 244–260.

Allen, S. G. (1985) 'Why construction industry productivity is declining', *The Review of Economics and Statistics*, 67 (4), 661–669.

Berg, S. A., Forsund, F. R. and Jansen, E. S. (1992) 'Malmquist indices of productivity growth during the deregulation of Norwegian banking, 1980–89', *Scandinavian Journal of Economics*, 94 (su), 211–228.

BERR (2008) *Industry performance report 2008*, Department of Business Enterprise and Regulatory Reform, London.

Charnes, A., Cooper, W. and Rhodes, E. (1978) 'Measuring the efficiency of decision making units', *European Journal of Operational Research*, 2 (6), 429–444.

Chau, K. W. and Wang, Y. S. (2005) 'An analysis of productivity growth in the construction industry: a non-parametric approach', in: Khosrowshahi, F. (ed.) *21st Annual ARCOM conference*, 7–9 September 2005, SOAS, University of London, Association of Researchers in Construction Management, 1, 159–169.

Chiang, Y. H., Li, J., Choi, T.N.Y. and Man, K. F. (2012) 'Comparing China Mainland and China Hong Kong contractors' productive efficiency: a DEA Malmquist Productivity Index approach', *Journal of Facilities Management*, 10 (3), 179–197.

Chiu, C. Y. and Wang, M. W. (2011) 'An integrated DEA based model to measuring financial performance of construction companies', *WSEAS Transactions on Business Economics*, 1 (8), 1–15.

Choy, C. F. (2008) Productive efficiency of Malaysian construction sector, *Proceedings of 14th Annual Conference of the Pacific Rim Real Estate Conference*, January 20–23, Kuala Lumpur. www.prres.net/papers/chia_productive_efficiency_of_malaysian_construction.pdf

Coelli, T., Prasada Rao, D. S. and Battese, G. E. (1998) *An introduction to efficiency and productivity analysis*, Kluwer: Boston.

Dacy, D. C. (1965) 'Productivity and price trends in construction since 1947', *Review of Economics and Statistics*, 47 (4), 406–411.

Edvardsen, D. F. (2005) 'Economic efficiency of contractors', in: Kaehkoenen, K. and Porkka, J. (eds) *Global perspectives on management and economics in the AEC sector, Volume II*, 11th Joint CIB International Symposium: Combining Forces – Advancing Facilities Management and Construction through Innovation, 13–16 June, Helsinki.

El-Mashaleh, M. S., Minchin, R. E. and O'Brien W. J. (2007) 'Management of construction firm performance using benchmarking', *Journal of Management in Engineering*, 23, 10–17.

Färe, R. Grosskopf, S. and Knox Lovell, C. A. (1985) *The measurement of efficiency of production*, Kluwer-Nijhoff: Boston.

Färe, R. Grosskopf, S. and Margaritis, D. (1996) 'Productivity growth' in: Silverstone, B., Bollard, A. and Lattimore, R. (eds) *A study of economic reform: the case of New Zealand*, Elsevier Science: New York.

Färe, R., Grosskopf, S., Lingren, B. and Roos, P. (1994) 'Productivity developments in Swedish hospitals: a Malmquist output index approach', in: Charnes, A., Cooper, W., Lewin, A. and Seiford, L. (eds) *Data envelopment analysis: theory, methodology and applications*, Kluwer Academic Publishers: Boston.

Färe, R. Grosskopf, S., Norris, M. and Zhang, Z. (1994) 'Productivity growth, technical progress and efficiency change in industrialized countries', *American Economic Review*, 84 (1), 66–83.

Färe, R., Grosskopf, S., Yaisawarng, S., Li., S. and Wang, Z. (1990) 'Productivity growth in Illinois electricity utilities', *Resources and Energy*, 12, 383–398.

Farrell, M. (1957) 'The measurement of productive efficiency', *Journal of the Royal Statistical Society*, Series A, CXX, 253–281.

Horta, I. M., Camanho, A. and Moreira da Costa, J. (2012) 'Performance assessment of construction companies: a study of factors promoting financial soundness and innovation in the industry', *International Journal of Production Economics*, 137 (1), 84–93.

Ingvaldsen, T. (2005) 'Scientific benchmarking of building projects – model and preliminary result', in: Kazi, A. S. (ed) *Systematic innovation in the management of construction projects and processes, Volume III*, 11th Joint CIB International Symposium: Combining Forces – Advancing Facilities Management and Construction through Innovation, 13–16 June, Helsinki.

Knox Lovell, C. A. and Schmidt, P. 1988, 'A comparison of alternative approaches to the measurement of productive efficiency', in: Dogramaci A. and Färe, R. (eds) *Applications of modern production theory: efficiency and productivity*, Kluwer: Boston.

Li, Y. and Liu, C. (2011) 'Construction capital productivity measurement using data envelopment analysis', *International Journal of Construction Management*, 11 (1), 49–61.

McCabe, B., Tran, V. and Ramani, J. (2005) 'Construction prequalification using data envelopment analysis', *Canadian Journal of Civil Engineering*, 32,183–193.

Price, C. W. and Weyman-Jones, T. (1996) 'Malmquist indices or productivity in the UK gas industry before and after privatisation', *Applied Economics*, 28, 29–39.

Rosefielde, S. and Mills, D. Q. (1979) 'Is construction technologically stagnant?' in Lange, J. E. and Mills, D. Q. (eds) *The construction industry*, D. C. Heath and Company: Lexington, MA.

Schriver, W. R. and Bowlby, R. L. (1985) 'Changes in productivity and composition of output in building construction, 1972–1982', *Review of Economics and Statistics*, **67** (2), 318–322.

Stokes, H. K. (1981) 'An examination of the productivity decline in the construction industry', *Review of Economics and Statistics*, **63** (4), 495–502.

Wang, H., Ye, G. and Yuan, H. (2010) 'An AHP/DEA methodology for assessing the productive efficiency in the construction industry', *Management and Service Science 2010*, International Conference, Wuhan, China, 24–26 August.

Wang, Y. S. (1998) *An analysis of the technical efficiency in Hong Kong's construction industry.* Unpublished PhD thesis, University of Hong Kong, Hong Kong.

Wang, Y. S. and Chau, K. W. (1997) 'An evaluation of the technical efficiency of construction industry in Hong Kong using the DEA approach', in: *Proceedings of ARCOM97 Conference*, Cambridge, 690–701.

Wang, Y. S. and Chau, K. W. (2001) 'An assessment of the technical efficiency of construction firms in Hong Kong', *International Journal of Construction Management*, **1** (1), 27–29.

Worthington, A. C. (1999) 'Malmquist indices of productivity change in Australian financial services', *Journal of International Financial Services*, **9**, 303–320.

Xue, X., Shen, Q., Wang, Y. and Lu, J. (2008) 'Measuring the productivity of the construction industry in China by using DEA-based Malmquist Productivity Indices', *Journal of Construction Engineering Management*, **134** (1), 64–71.

13 Endnote

Rick Best and Jim Meikle

Measuring and comparing things has become increasingly important in recent years. We have league tables and key performance indicators for all aspects of society and the economy. But often it appears that we are not necessarily getting better at measuring and comparing, nor do these exercises necessarily lead to improved outcomes.

In the preceding chapters, the various authors have explored a number of aspects of the broad topic of 'measuring construction'. They have looked particularly at measurements (and comparisons) of construction cost, value, performance and productivity (at several levels: site, project and industry) using a variety of tools and techniques. These are all interconnected and are to a large extent quantitative; they could, therefore, be expected to be more tractable than, for example, more obviously subjective issues such as construction quality. It has been demonstrated, however, that this is not the case.

Much that is written about construction prices, output, productivity and the rest is unreliable or plain wrong, sometimes deliberately, often not. It is hoped that some of the approaches and methods described and discussed here will help future analysts and policy makers. Unreliable advice based on unreliable data more often than not leads to unreliable policy.

The process of writing and editing this book has helped the editors clarify some issues but also identify gaps in their knowledge and understanding. It is apparent that there is still a good deal of work that remains to be done before mature methods that are generally accepted as providing reliable and robust results are identified, tested, developed and in use. More work is needed on both data and methods and, hopefully, some of what is written here will stimulate thoughts from readers. If it does, the editors would like to know.

Over the years, attempts have been made to measure and compare both construction projects and whole industries. A variety of proxies for construction have been used as researchers have looked for ways to make comparisons that not only are valid but can be made without the need to invest heavily in complex data collection. It would be good to think that we may one day stumble upon a single, easily measured and/or priced item that will reliably reflect construction as a whole. It is unlikely, however, that such an item will be identified; for the same reasons that the Big Mac hamburger is not really representative of all consumption in an economy, it is doubtful that any simple item will accurately reflect, for example, construction price differences overall. In truth, the heterogeneity of construction output even makes use of the term 'construction price differences', a dubious proposition as the cost to build a warehouse or a school may have little connection to infrastructure prices even in developed economies and even less to do with the cost of constructing a dwelling, particularly where a large proportion of the population live in vernacular structures made from local materials such as wattle and daub or mud brick.

A major issue with virtually all of the topics covered in this book is that we don't know when we have the correct answers. We know that commercial exchange rates are not much use for comparing construction price levels between very different countries and that general price indices are not much use for measuring construction price trends within a country. Purchasing power parities (PPPs) for construction work and construction price indices are better, but they are broad indicators; they don't necessarily apply to particular projects or particular types of work. More work is required to develop and refine techniques that will help produce more reliable results at greater levels of detail.

Work is needed on weights and weighting. Expenditure weights (i.e. the ratio of expenditures in different sectors of construction, typically but certainly not always categorized as residential, nonresidential and civil engineering) are probably the most straightforward, although they will vary over time, particularly in smaller and less-developed countries. Resource mixes (i.e. the ratio of labour, materials and other inputs), however, present a rich topic for further research. The International Comparison Program (ICP) has collected mixes for around 100 countries, and there are a few published references, but more effort is required to establish how and why mixes vary across countries, over time and by type of work.

Weighting of items within resource groups such as labour and materials is important, but the use of unweighted (or perhaps more correctly, equally weighted) price relatives can provide an acceptable alternative to quantity- or value-weighted prices for groups of similar items. It is important, however, to select appropriate items and to include the right number of items in whatever we use to represent construction sectors or construction as a whole.

Definitions of construction output or activity vary. The tendency is to assume that it includes all – or most – construction contracting activity, but it is less clear what is included in respect of construction professional services, self-build construction, do-it-yourself (DIY) construction activity and 'black' or 'grey' construction. There is plenty to be done, for example, if we are to find ways to reliably include the value of informal construction in national measures of construction activity; at present, particularly in developing economies, such work can go largely unrecorded and may represent anywhere from 50% to 90% of all work done.

There are lessons to be learned for spatial price indices from temporal price index methodologies and *vice versa*. PPPs are an international issue and are dealt with internationally, while price indices are largely national concerns. It does seem, however, that if data is being compiled regularly for PPPs – as part of an international obligation – the same data could be used to measure price changes over time. In fact, it may be more appropriate for national price index data to be recycled to produce PPPs. One benefit of this could be an increase in the frequency of generation of PPPs; at present, the ICP produces PPPs some years apart, and when the latest PPPs are published, they are generally well out of date: PPPs based on the 2011 pricing round did not appear until well into 2014. Another benefit could be standardizing price index methods internationally. Common or overlapping data sets and standardized methods would help to reduce potential duplication of effort. Wealthy developed countries can support large national statistical efforts but smaller, poorer nations have understandably different priorities when they are allocating their meagre resources. It can be truly said that 'you can't eat statistics'.

In addition to methods, data is the other key component – without sound data, the best tools and methods are of little use as a means of measuring construction in much the same way that a Rolls Royce without fuel is of little use as a means of transport. Data issues relate not only to the availability of data but also to the uniformity in the way data is recorded and presented, and both aspects probably require better understanding and appreciation of the value of good data by governments and their agencies. Only if that is understood will policy develop that drives the collection and standardization of construction data. If data availability and quality improves and we can continue to develop more refined measurement and analysis tools, then we may eventually be able to produce more robust measurements and comparisons of construction around the world.

Different users of construction data have different needs. For example, the needs of the construction industry and national statistical authorities are very different. The construction industry is most interested in data on specific projects or project types in specific locations or areas at specific points in time; national statisticians want to measure or compare broad groups of economic activity, and they are interested in national averages or quarterly averages. It is also important that different users understand each other's needs.

Different users will also use different methods, and methods may be chosen depending on the data required to fuel them. The ICP now surveys around 180 countries in gathering data for the production of PPPs, and the methods they use must be practical for countries from Albania to Zambia. If we use the Rolls Royce analogy again, we can say that the most excellent method for making comparisons is not very useful if the users can't afford to obtain fuel (data) to make it operate. While practicality of implementation across a broad range of countries is important to the ICP, researchers attempting to make comparisons between wealthy, developed countries such as the UK and the US can use more sophisticated methods that require specific data collection and, if necessary, the normalization of data that is inconsistent. In truth, there may be little point in trying to compare industries between countries as disparate as, say, Liberia and Australia; the results of such exercises would most likely be of little practical use, whereas the nature of the ICP and its aims mean that as many countries as possible, and ideally all countries should be able to contribute without excessive cost.

This book shows clearly that how we should measure different facets of the complex industry sector that is construction remains open for debate and development. While we acknowledge that there is no 'correct' answer to any of the questions that this book explores and that will always remain a complicating factor, whatever we do, we can only hope that we are incrementally improving the way we measure and compare prices, productivity and so on. It is perhaps only by applying a variety of techniques to the various problems and comparing the results that we obtain that we will know if we are getting closer to developing an acceptable set of tools and methods.

Index

Note: Page numbers followed by *f* indicate a figure on the corresponding page.
Page numbers followed by *t* indicate a table on the corresponding page.

Printed in the United States
by Baker & Taylor Publisher Services